THE
ROAD
TO
FOX HOLLOW

The Road to Fox Hollow is published under Voyage, a sectionalized division under Di Angelo Publications, Inc.

Voyage is an imprint of Di Angelo Publications.
Copyright 2022.
All rights reserved.
Printed in the United States of America.

Di Angelo Publications
4265 San Felipe #1100
Houston, Texas 77027

Library of Congress
The Road to Fox Hollow
First Edition
ISBN: 978-1-955690-07-2

Words: W.C. Leikam
Cover Design: Savina Deianova
Interior Design: Kimberly James
Editors: Cody Wootton, Stephanie Yoxen

Downloadable via Kindle, iBooks, NOOK, and Google Play.

For educational, business, and bulk orders, contact sales@diangelopublications.com.

1. Biography & Autobiography --- Nature
2. Nature --- Animals --- Wildlife
3. Science --- Life Sciences --- Zoology --- Mammals

THE
ROAD
TO
FOX HOLLOW

W.C. LEIKAM

CONTENTS

PREFACE

I began this book because it rather urged itself on me as it rumbled from within and onto the page. This began as an essay: my attempt to organize the wild. I thought I knew about the gray fox and was arrogant enough to think that someone else might enjoy such a story—that someone may just awaken, as I had. The more I wrote, the more I doubted, and the more poetic it became.

Much of the science surrounding foxes—and most animals, for that matter—came from the 1950s, and the definitions and classifications that they asserted then are still in effect today. As I probed the scientific, peer-reviewed papers on the subject of the gray fox (known there as *Urocyon cinereoargenteus*), I saw huge gaps between what I witnessed out in the early morning darkness before dawn and what the scientists had to say. As time went by, I learned that some of the things people thought they understood about the gray fox were clearly false—information that had been handed down generation after generation without any scrutiny. No one cared to question the "facts." No one stopped to double-check. I saw that the doors to understanding the gray fox were locked by tradition, barricaded by so-called "objectivity." Somehow, I needed to unlock those doors and show these foxes as they truly were.

Before I make any assertion, I require that three different foxes, preferably living some distance apart, exhibit the same or similar behavior on three separate occasions. Even after that, I am skeptical, ready to alter my ideas for something more definitive. Exceptions abound, causing a philosophical and pragmatic tension within, a furious conflict that drives me. I must sort it out. Curiosity drives me.

The gray fox in central California falls under the subset known as *"townsendi,"* and this is the type of fox I've come to know so well. They came into my life just as I came into theirs. Townsendi lives at this story's core, and I knew in my heart from the onset that there were two hard truths I had to live by in its telling. First,

I could never objectify the foxes as the scientists had, regarding them as faceless specimens or raw data. Second, no matter how much I tried, I could not keep myself out of this story. This book is about foxes, and I am a man, so what business have I in discussing myself when the foxes deserve center stage? It is because I saw them through my own eyes, and for that reason alone, I must include myself in the tale. After all, I dreamed of them, thought of them, and lived through them for all these twelve exciting years. I saw their history play out before me day after day, in the dark of early morning before dawn drew the Diablo Mountains to the east against the sky. I was there, I am there still, and in some ways, I hope to always be there among them.

Curiosity and, unfortunately, some arrogance drove me to study the foxes, to observe and sample and collate their lives with my superior intellect. What I discovered was that they taught me. They were the instructors and I was a hapless pupil. All my knowledge was nothing more than smoke in a warm breeze.

Here I will take you on a journey into the thicket, where the breeze rises to a gust and the trees chop in the gale. The story changes ceaselessly because its designs, its vibrations, and its tones instinctually cannot rest. As it unfolds, so might we. The tale of the gray fox must be told, because perhaps it will give us insight into who we are, or grant some sense of balance, or lend us the measure by which we govern our lives side by side with all bonded lifeforms. It may even allow us to recognize our compassion rippling within and embrace the wild instead of destroying it. Finally, this book is nonlinear in that it meanders through time just as the gray fox moves through its home range.

This narrative rests on a bedrock of truth, woven together with the heat of conviction from my own observations. It is a personal story of thought and patience in battling my own ignorance, and this human's pursuit of an open mind. Of necessity, I take sketchy notes when in the field—the kind that jogs the memory back at the computer. I've developed a shorthand that allows me to contain all my observations from the morning on a single Post-it. To tell this story, I must draw from that abbreviated log and remember the details to the best of my ability. I have no way of knowing a fox's motivations, and so I must speculate through the art of quasi-fiction as I give flesh and bone to the inner workings of these animals. It is as true as my limitations allow.

This account needs other witnesses, documentarians, and researchers to corroborate or challenge my observations. What follows is not to be considered the final word on gray foxes, but only a beginning. If any notion, assertion, or observation that I suggest here stirs biologists or PhD grad students to double-check, to replicate, to challenge what I submit within these pages as true or false, I encourage them to do so. I urge those with the tools to step forward.

This book is the work of one person's studied perspective, and within that perspective I automatically project my biases as a human being. I cannot be known as holding the final truth about the behavior of the gray fox, because such a thing may not exist. I would like nothing more than to know that my impressions were confirmed through good science, but I am likewise honored when proven wrong. Failure is a wonderful professor.

ONE

THEY'LL TRAP YOU

Gradually, they altered my life. I slid unconsciously through time, completely unaware that these foxes crafted my days. They changed me like none other, save only my ex-wife. My first impactful encounter with the gray fox years ago, at thirteen or fourteen, happened when I carried that fox home in my arms. It began the long thread between my youth and old age.

To us as children, these foxes were incomprehensible beyond the commodity they afforded. We could use the extra money we'd earn for their capture: seven dollars a pelt and maybe ten for a live one. For two kids living down in Freedom, California all those years ago, that was a fortune.

And besides, what was the life of a few varmints compared to cold, hard cash? They had no character, no language, no future. How could they be anything but wild animals running headlong into the inevitable?

"Every little bit counts," Mom always told us. As a Dust Bowl family, we'd narrowly escaped poverty because Dad was smart. He could look at a problem and know the solution right away, and though he worked himself down to the bone, we still knew the pressures of want. His own long road eventually led to managing the apple cider vinegar plant in town, owned by the Speas Company out of Missouri. He was the boss, and I respected him as only a son can. I wanted him to be as proud of me as I was of him.

The feed store sold traps of all sizes, and we stood there, my brother and I, staring at the wall of teeth, chains, coils, and stakes with a sense of dreadful mastery. Those implements of pain and terror gave us power over all living things and affirmed our right to rule—provided they didn't snap off our hands in the process.

"Wanna fox, huh?" said the man at the counter. We bobbed our heads up and down. "Ya find they trail an' put this here 'longsides." He pointed to a trap hanging above us. "Seen some few up by Diehl's land, in that floodplain upstream." I lifted the trap from its hook and brought it to him, sliding it over the counter as I laid two dollars in his palm.

"Snatch you some dog food an' sprinkle it all 'round yer trap when it's set," he added as we made for the door. "Gits 'em sure as shit."

We looked to the brush for signs and found an animal trail, then crawled on all fours into the thicket, steel trap clanking behind. When we found a likely place, we staked down the chain. I dumped a sack of dog food on the ground and scattered it with my foot. We set the trap, prying apart steel with what shaky strength we could muster against the taught springs. The trap seemed just strong enough to hold until we could return the next morning before school to see if everything had worked.

In the early light, caught at the leg and still as stone, was our four-legged payday.

One trap wasn't enough for long. Greed led us to hunt glass bottles in the grass along the roads, and the five or ten cents we got for redeeming them at Freedom Market all went to buying more traps. We set five traps at a time and checked them every morning. Gone was the impulse to help Mom and Dad with the monthly bills; our only thought was getting money to get traps to get foxes to get money.

We didn't sell them all; sometimes the traps were

more gentle, and we'd stuff the foxes down in a gunnysack. Scrounging together enough wood and wire mesh, we built a pen to hold a few, with the idea being to contain them until the market price improved. Their little eyes behind the enclosure wire showed that they were lost and concerned, but not afraid.

One afternoon, my brother and I went along the creek bed to check our traps, strung far and wide upstream from home. I heard the chain rattling before we even got close.

"We got one," I said. "Get that sack ready."

My brother readied himself as I leaned down to grip the fox by the scruff of its neck, but the instant my hand touched its fur, the fox slumped into utterly fluid docility. When the trap fell from its leg, the fox squirmed a bit but didn't try to bite me or get away. I couldn't tell if it was frozen with fear or was grateful for the help, but the bag suddenly wasn't necessary. *Something* happened then within me, and I scooped the gray fox up in my arms to cradle it like a baby...and it allowed me. Holding the fox against my body, I headed home with my bewildered brother in tow.

We came up through the rentals that Mr. Johnson owned, and a woman who lived there was out in her yard. Even now, I can see it all unfold. I know exactly where her house stood. She saw us coming and moved to intercept. She came out to the dirt road with all the severity of old age and snapped, "Whacha got there?" She shook her finger and pointed at the fox.

"It's a fox, ma'am," I said. "A gray fox."

She frowned, then asked, "A wild fox?" I hesitated, then nodded. Her brow furrowed as she scowled at it.

"Gonna bite ya, ya know. Bet mah bottom dollah. Careful, hear?" Her eyes flared with fear and she shot her finger at me. "Could have rabies, ya know. Ain't natural, you holdin' a animal like that. Ain't natural, I say. Spooks."

I worried for an instant that she might be right, but then I looked down at the creature in my arms and knew with certainty that her fears were unfounded, outdated. "Oh, it's okay," I told her. "He's fine." I stepped around her and went home to put the fox in our pen with the others.

The next day, we arrived at the pen to see a hole dug beneath the boards we'd buried a foot deep to keep the foxes in. They'd escaped, which baffled us in our arrogance. I remember noting, *Those little guys must dig fast to get out in one night.*

We modified the pen to keep them cooped up, but no matter how many hours of sweat we poured into it,

17

no matter how ingenious we thought ourselves, they always escaped again. Despite our "higher" intellects, our tools, and our materials, we could not match their persistence and will to be free. As long as they were alive, they would win.

Dad watched. He didn't interfere, nor did he help us solve our problem. Maybe this was his way of testing us, but I wanted his approval so badly that it twanged in my guts. I yearned to trap a whole pen full of foxes, to march Dad down to the cage and show him how capable I was. In my mind, he'd see my handiwork and give me a proud pat on the back.

I never got it.

"Be careful," he told me one night at supper. "You keep gettin' those foxes like that and one of these days they'll trap you."

It was about 4:30 in the afternoon when my brother and I headed up the hill to the main road where the apple trucks came through, loaded with boxes of tart Pippins or juicy Bellflower apples fresh from the tree, destined for places unknown to us. We ate them at our leisure. We were headed on up toward the Green Valley Bridge when I looked over, and there, reaching for a low-hanging apple, was a gray fox.

He was one we'd trapped and caged—and later lost— and I recognized him by his severe limp: I'd accidentally broken his hind leg while trying to pull him out of a trap, when he'd leaped out of my hands and cracked the bone. He held that paw off the ground as he made for the apple. The bandages and my makeshift splint, which I'd fastened around the break before putting him in the pen, were gone. He suddenly jerked in pain, unable to reach the apple, and I realized the suffering my error had caused him. With a twitch, he fled, limping through the brush on the search for easier food.

I remember hoping that he would be okay and maybe eventually regain the use of his leg. Such wishful thinking for a child. I know the unfortunate truth now—that an injured fox's chances of survival are next to nothing, and that I'd doomed that creature to a slow death by starvation.

We didn't know what we were doing, and our ignorance was causing irreparable harm to a group of animals we should have left alone.

When all was said and done, we pulled our traps from the woods along the Corralitos Creek and let Mom fill our fox pen with chickens. Half a century later, another fox would strike a chord with me. I'd feel that deep pang of sadness drenched in curiosity tug at me. It was heavy as a teenager, and heavier then as a man. I can still see that gray fox down there beneath one of the apple trees in Old Man Resitar's Orchard, reaching for a need that would never be filled.

Dr. Ben Sacks at UC Davis wrote in a letter concerning my work that the gray fox developed from "the most evolutionarily basal canid, representing a more ancient lineage than wolves, coyotes, jackals, the South American canids, and all other foxes, including the odd bat-eared foxes of Africa and raccoon dogs of Eurasia. Despite their key position in the tree of life, there have been no detailed studies to date of the behavioral ecology of gray foxes, other than to document home ranges and habitat use based on radio telemetry." Dr. Sacks has pinned the date of the emergence of the gray fox to between eight and twelve million years ago, originating in what is now California.[1]

That Corralitos Creek is a living thing. It's still there,

of course, and will be there long after all of us are gone and this story is forgotten. The creek has supported all manner of life just beyond our reckoning, whether dry with sand and stones or flowing deeply when the rains come. If we had two, maybe three, inches overnight, we stood at the kitchen window watching the brown, boiling water rush downstream through our backyard. It was a haven for us kids and, at the same time, a looming threat so very close at hand. So it is with humans: no matter where we might be, anywhere on the planet, we are always drawn to water, our enduring holy grail.

The gray fox is so ancient that it has withstood global warming and ice age episodes repeatedly, and it continues to thrive. It has known periods of thousands of years of hardship when the great ice sheet gouged out the Great Lakes. What does it know that allows it to survive so well? There must be something within this small canid's mind that can teach us how to manage and survive the warming cycle we see everywhere. Could following the gray fox teach us how to survive, too?

More than fifty years have passed since my brother and I pulled up our traps. The habitat of the gray fox now lies destroyed, overtaken by an invasive species. Houses stand on the creek's banks where foxes and raccoons and other night-prowling mammals once roamed and hunted, denned up, and birthed their young. They understood the ecology of an optimal people-scape and lived their lives as best they could under the circumstances. It could be that they still prowl there in the darkness between trash cans and boat trailers, but reality—humanity—says otherwise.

Their voices have been silenced and their cultures erased from time by driveways, sprinklers, barbecue grills, swing sets, and patio furniture.

What is the effect of a mass extinction on those wild cultures that never understood what was happening? They didn't know where the others of their kind went, but one day they were simply gone. Pertaining to foxes, I call this "The Vanishing." What might it feel like if you knew that you were the last of your kind? How would that affect humanity, to feel the distinct lessening of one's species rather than the constant, overwhelming increase?

It behooves us to seldom intercede in nature. The forces of evolution have brought us to this moment in time. Look at our long history of unnecessary destruction: we allowed our egos to govern without understanding all the dynamics of nature. What might happen if we stopped trying to control every inch of this planet?

We cannot call ourselves civilized until we freely give all living things on planet Earth their rightful place alongside us. The best practice is to allow nature to live for itself, and for us as humans to find a comfortable niche in that greater universe of being. This may very well bring about the eventual death of the city, that needlessly large-scale social monstrosity, which I can understand is a frightening concept for most. Will we ever return to the balance where we understand that we are multifaceted mammals feeding in the primordial soup of time, the same as every lifeform on the planet? We share the same bowl. We do not own it. When will we accept that all life is equal? It is absurd to think otherwise.

TWO

GENESIS

In the beginning, I set up camp in what I'd eventually call "Fox Hollow." It wasn't an ordinary camp—no sleeping bags, no air mattresses, nothing but the cold earth. I didn't stay there overnight (I had a lot to learn). Instead, I drove down there every morning in my noisy pickup and set up my post just across the road from their den, my back to the salt marsh. That's where my observation of the gray fox began, down on Landings Road. That's where I catalogued a gray fox family with photographs, Post-its, and moments of pure, breathless wonder.

So much happened down in the big clearing beneath the canopy, there at the crossroads. Runners and joggers moved along, walkers with dogs on leashes, even workers from the high-tech center nearby. Those who passed by were ignorant to the drama, the play, and the learning that took place just there beneath the canopy, no more than ten feet away.

From my vantage point, I saw the pups—I use the term "pups" instead of "kits," as they are canids— playing under the canopy. In the area I referred to as their "front porch," the pups learned to become adults; they even practiced having sex with one another.

Before the male could deliver dinner (had evening come so quickly?), a pup pushed its way through the brush, paused, scanned me with its ears, and came a few feet toward me. The camera's shutter chattered. Slowly, very slowly, I stood to get a good shot. The pup vanished. It was like a spring letting go. There one moment, gone the next. I'd never seen anything move that quickly. These foxes were wild and skittish, or so I thought—as I've said, I had a lot to learn. In time, I learned not to move my feet. I could move my hands; the foxes didn't feel threatened by that, but if I shifted my feet, they vanished.

I wondered, *What would it feel like to have a fox touch me and I touch it? How would that feel? Emotionally, I mean. What might happen?*

It was that firstborn, of a litter of two there in Fox Hollow, that paved the way for that to happen. Seems as though, no matter which natal den, there are always one or two pups who brave the unknown. They are fearless. The others of the litter cower and look on.

The pup returned later, curious about my presence, and edged slowly closer to where I sat. I reminded myself to stay still, and was rewarded with his full interest. The pup came to me and sniffed my boots. I thought, *What if he licked my fingers?* That young fox came closer, then hesitated as I extended my fingers and urged, "Hey, little fox," in the softest voice I could. Caution rippled through him as he wrestled with his fears. His nose wavered, but then, with tentative care, touched my outstretched fingers. Sheer magic touched me, connected with me, crying *"Reunion"* with the wild. He licked my fingers and I sat there unable to move, absolutely stunned.

When that pup, the one who I came to call Squat, sniffed and reached toward me with his shiny, wet nose and touched my fingertips, I felt a subtle transfer—a hum—between us. Instantly, I understood something that, two seconds before that moment, I could never have conceived. The wet of that black nose shifted something within me that I still feel today, ineffable yet undeniable.

When my thoughts returned, I knew that it was as a bond ignited. I shook. I can't explain it. Ever since that moment, he lives within me. There are stories to be told about him—stories that no one has time to hear—and now he is gone.

I reached to pet him, but he dashed away so quickly that I couldn't touch him. The flight response in a gray fox is powerful. Muscles flash to action long before thought. The only way to delay the escape is through tempting their curiosity.

After that singular encounter, Squat returned to his skittish nature. As long as I kept taking pictures or stood still, he came close, but if I even slowly raised my hand to scratch an itch, he was gone in an instant. His response was the wild announcing that it was alive, alert, and well. I saw him come out to look at me and, when I stepped to the side, he ran back to the edge of the brush, turned, and lay there, chin on the grass, looking back to wonder about that human out there.

Squat unknowingly became my first teacher. His parents were disconnected, and he lived on the border between high-tech medical laboratories that were pioneering the next medical instrument that could change the way physicians did their jobs and the international law firms whose power and history ran back into World War II. Their heritage is nothing compared with Squat's: that gray fox is the basal canid at ten million years old.

They spoke to me; they beckoned. I don't know why I listened, but I couldn't deny them. Was it me seeking atonement for what I had done to the gray foxes down along the Corralitos Creek as an innocent kid? My Buddhist brother, Lee, says that I am burning my karma.

It was wild bush dogs that caused such an insane pull in my gut that the drive to know—to understand this alien species—could not be dissuaded. I could not shove this aside and let it go. It clung like sticky clay. I didn't know a thing about them, and in the long haul, there were no scientific papers to confirm what I was actually seeing.

So I stepped out and made myself known to these foxes. I sensed that if I was going to actually understand their behavior, I needed to burrow my way into their inner workings, into their family life, to catch that intimate glance I sought.

I had no idea how to accomplish that strange feat. I mulled it over and over again; nothing jelled. In the beginning, I'd seen that they were as curious about me as I was about them. I kept stumbling around trying to ferret out their behaviors. I found myself thinking of them, wondering, hungry to know them in even more depth, but it seemed that day after day I stood in the same spot, making no headway in my understanding.

I had been watching them down in the Hollow—a dip in the road where I was able to sit day by day—when one morning, no foxes appeared across the road. I thought it odd, but hope drew me back. They were absent the following day, and the next, and the next. I sat watching the brush on the other side of the road, hoping beyond hope, tense, anticipating that the brush would rustle and that a fox's face would poke through the bushes, but nothing. Silence.

Something had certainly happened. Had my presence driven them away, or was there some more sinister force at work? A week passed, then another. I remember waiting at length and realizing, quite suddenly, that they were gone. The knowledge of this fact gnawed at my bones for a time, but I discovered I could let it go. I sensed somewhere deep in the pit of my being that I had a very long way to go.

My journey hadn't quite yet begun.

THREE

THE GRAY FOXES
OF THE BAYLANDS

I've seen these foxes in so many varied situations. I've seen the relationships between them and how, on occasions, gray fox friends hunt and even fight together. I know who gets along with whom. Tippy's been here from the beginning, or shortly thereafter. I named her Tippy because, when she was young and came out into the big clearing to look up at me, she often tipped her head from side to side, as though trying to figure out some complex problem. The problem, as it turns out, was me.

We understood each other, which was why Tippy came and licked my hand if I extended it, and why she sank her sharp, tiny claws into my leg. Tippy trusted, but if I ever brought a new observer with me, she kept off to the edge of the brush, watching. More than the rest, she was acutely alert to her surroundings. She measured her tolerance.

The alpha male, Dark, ruled Tippy's home region and had done so for more than a year before I dropped into the scene and changed everything. But one night, he sniffed the air and smelled the scent of an unfamiliar female approaching. He already had both ears pointing in her direction. He waited. She came out into the big clearing from back on the thicket trail. She knew that he was there.

He raised his tail in an arch, hunched his back, laid his ears flat back on his skull, and extended his legs to make himself look more fearsome than life itself. She didn't move. She watched. She tensed slightly, for she read that battle pose that he'd just struck, but she didn't believe that he would charge her and fight.

He didn't move. She walked over next to the black walnut tree. He followed, came up to her, and sniffed beneath her tail. She crouched in that submissive way that only happens when the female is on the verge of estrus. He came around to face her, sensed that she had relaxed, and as she stood, he nuzzled and groomed her flank.

She accepted him. I named her Cute, and off she went to hunt with Dark. She heard a woodrat padding along a branch. Dark saw that she'd cocked her hind legs in preparation of vaulting, that arching leap that would come any second. She bounded. She emerged with a fat woodrat, tore it apart, and did not share a nibble with the others. To share is not the way of the gray fox. To share does not exist in any of their layers of feeling, emotion, or cognition.

Dark caught his own meal nearly an hour later. He tucked away the forequarters of the squirrel for a snack. Cute decided that she would hang around. Maybe Dark was the mate she'd been looking for. She saw that he was able to hunt with an uncanny skill. If they had a litter, she knew he would provide, so Dark was hers.

In the background, officialdom banged on my window. They wanted a piece of whatever I was doing out there on what they called "their open spaces," as if anyone could own the Earth. The city shoved papers at me to be filled out: agreements, contracts, and permits. I shelled out $300 here, another $250 there, paying various agencies in the city and the state, buying permission to do what I'd already been doing. They charged for anything they could dream up, because this was new territory for all of us. I often asked myself where the money went, but I didn't have time to find those answers. The real work was far too important.

Why hasn't someone else looked deeply into the lives of all these urban animals living beside us, next door, around the corner, around the planet that we mutually inhabit? Why haven't we become neighbors instead of adversaries? The rift between us is real, but what purpose does it serve? How is it possible that we can live as neighbors and not know a thing about one another? We alienate ourselves. Neighbors no longer know neighbors, and they turn a hateful gaze upon the wildlife crossing our lawns in the dark to stop on our porches for a bite to eat.

Information is dead. It's heresy to consider, but might it be that science, the method that gives us knowledge, inherently contains severe limitations in the way it frames itself? Is objective observation fallible, in that it cannot see the whole picture? If so, are such limitations only presenting us with a fragment of the truth, only a snippet of these foxes' reality, only a glimpse into their being? Of course, I believe so.

When, or if, we decipher their language, whatever form that might take, may it be a different reality illuminated from beneath the shroud of ignorance.

I carried my camera and took photos of my obsession. I admit it freely, because that's what the gray fox was to me: an obsession. I walked closer, my camera capturing the gray fox sitting there until I got right up to the gate. That little fox was just on the other side, perhaps a dozen feet away, and it hadn't so much as twitched a muscle. *Odd*, I thought. Leaning against the gate's post, I froze with sudden fear. *Rabies?* A chill ran through me. I looked for the signs. Nothing around the mouth, no shaking, no loss of balance, no outward aggression. *Looks okay...* Doubt hung there, nonetheless.

Recklessly, I moved around the end of the gate and faced off with that wild gray fox, but it just turned away and dissolved with the brush. My camera's shutter hammered and clattered away.

The next day there was nothing in the dip of the road that linked Landings Road to Harbor Way. Because that fox had been there before, I returned for several consecutive days, each time fearing I would hear that stern voice behind me, commanding that I STOP, that I had trespassed and would have to pay some grotesque fine—a truly grotesque fine for walking there near a fox.

That voice would come years later.

One morning, while still on the hunt for that peculiar gray fox, I decided that I no longer wanted to sit in the middle of the road, waiting for the foxes to come back to the den or come out and make an appearance. The road was damp and cold—three feet above sea level. I needed something to sit upon. Up the road, a jumble of branches and stumps were strewn about. Someone had come through there with a chainsaw. I found a log that leaned a little, but was perfect as a stool to sit on and photograph these wonderful animals—these little bush dogs, as I called them.

I returned and there was that fox again. It came from the brush out onto a bed of grass where it lay on display. Shortly, a second fox emerged and sat there across the road from me. I didn't think; I just watched. Pieces of the puzzle flowed into a gestalt, merging understanding with sheer ignorance. I needed to know the depths of my ignorance, so I followed my curiosity. I knew that by letting myself go down that road, I might just discover something. I hungered to know more.

I soon understood that these were not two lone foxes, but a family living in that thicket across the dirt road. A typical urban gray fox family consists of a female, a male, and one to seven pups. Another version of the family adds a "helper" female. As her title suggests, she helps the family survive. She hunts for herself and the others, especially the pups, and even has a role in rearing them. In many ways, she's just like a nanny. This family didn't have a nanny, but I intended to keep my eyes open.

That first month with the family, it was like they wanted to perform for me. When I found them, it was right at the cusp before they would disperse, so I only got a few weeks before they vanished. I watched the pups grow into late adolescents, able to hunt for themselves. The adults no longer had to bring back jackrabbits, mallards, or young Canada geese. It became difficult to discern adults from pups. They played and romped and made mock sex, all in preparation of going out to find their own mates, their own places to sleep during the day and hunt at night.

It is said in the literature that males may disperse far from their natal dens, while the females stay closer to home and, when the situation is just right, have their litters in the same den where they were born. Later, I would learn about the need for gray foxes to do battle for their natal den. As a juvenile, the male I called Squat fought his parents for that den. I didn't see the actual battle there on the old dirt road; I saw only the outcome.

I have the date tucked away somewhere. The last morning before they all dispersed, I sat on my log stool, waiting. Dawn had already broken the night's dark back, cracked just a bit by spraying a dash of gray across the marsh and pink everywhere over the baylands. A ground fog hung there, so thick I couldn't see my boots beneath me. Sometime in that dim light I saw the ghost of a shadow emerge from the brush. My eyes had long been adjusted to the early morning darkness and I knew by its movement that it was one of the local gray foxes.

I talked to the foxes, knowing that they could not understand my words, but could understand the emotion behind them. "Well, there you are," I'd say as one came from the darkness. "Have a good hunt? Belly full? Ready to settle down and get some sleep?" Often, I chattered on just so they could hear the tenor of my voice. The little fox walked around me, went over by the bunch of grass at the roadside, and then came back, listening, ears scanning. A loose estimate would say that's about 80 percent of a fox's life.

The sun lay not far beyond the eastern horizon. Light etched its way into nooks and corners that had, moments before, been invisible. In the distance, the landfill came into being, squatting in its rut with the birds already circling overhead. That's where I expected our drama would unfold. In that growing morning light, my eyes played silver tricks on me. Phantoms emerged from the darkness down that old dirt road. Deep within Fox Hollow, something stirred.

Something went terribly wrong with the foxes, or so I think. The cause may have been long before we stopped connecting, running silent and deep across eons.

As we head back into the big clearing, we top the berm where the coyote bush grows maybe ten feet tall there on our left, where the foxes love to lay in the shade on hot summer evenings. It is a gathering place year round. From there, we slide down a muddy bank to the water channel below. When the creek floods, this channel fills with water. At its deepest it's maybe six inches. It has been raining an inch every couple of days. Everything is wet except beneath the big palm tree in the thicket.

The seasons course through and change mud to dust.

The gray foxes follow me as I follow them.

Tense, the nearly two-year-old female, lives in constant tension. She ceaselessly checks her environment, ensuring there are no threats dashing in from afar. This is her life. For instance, no matter how hungry she might be, before ripping flesh, she checks to make sure that there are no other foxes, crows, or hawks around to steal her catch. She is ever ready to fight to keep it, or pick up the rabbit and vanish into the thicket for a private meal. When she looks up at me, I see that she is searching for something; she works to understand me. I reflect her posture and we share a moment together as a fox and a man.

Big Guy is Cute's pup, the youngest. As I come up the channel, he is lying over near the edge between the open field and the thicket behind. I call, and he stretches once, then bounds down across the grass. He is jaunting, happy and apart from the rest.

Dark Eyes, the adult alpha female, reflects elegance and a simmering kind of serenity. She is tender and caring, maternal but with paternal strength. Whereas most of the other female foxes express themselves through their hypertensive demeanor, Dark Eyes has a placid, soft feel about her, one of being aware in the full sense of that word but at the same time smooth and gentle. Spiritual. When others run about, hiss, or growl at one another, she goes off by herself and lies down, apart from such a spectacle, watching the others expend useless energy over each other or over a carcass. She notes my presence, at peace.

Never have I seen her in a fight with others. Only once have I heard her growl deep in her throat, even in situations that may have called for it. She moves with a graceful, gifting gentility. Always she looks directly into my eyes. When I named her, the fur pattern around her eyes was far darker than most foxes'. The color was just shy of being a mask. She exudes the presence of being whole, of being complete, as though she'd accepted herself as who she was and had no desire, conscious or unconscious, to reshape herself or grasp for more.

She is satisfied within, and that's the way Dark Eyes makes me feel. She is the alpha female of the region— the queen, if you will. All of the other foxes approach low to the ground when they come upon her in the thicket, or under the canopy in the big clearing, or out on the overflow channel: ears flat back on their heads, slithering, tails swishing—the ancient mark of happiness—chirping bird-like sounds, approaching from under Dark Eyes' chin in full submission. They come to her one after the other. She sits there in her regal pose, silent and unmoved. She became the mystic, my Buddha dog, by which I mean she instilled a lasting peace within me.

These foxes gather and lie there in the grass all around. One scratches, one grooms. Over there, beyond the circle, Cute curled up a few minutes ago and went to sleep. I haven't seen her mate, Dark, for several weeks now. He must be off on some journey, or so I hope.

FOUR

THIS IS MY RELIGION

No longer might we simply get a scientific fragment of data and almost immediately make further assumptions. Those numbers do not call out with a story that leads to understanding. Though useful, that scientific temperament may not have the last word when it comes to gaining knowledge. We need to conceive of our own role in this web of life.

In this holistic paradigm that I propose, we can link all of the science that we can bring to bear, but we must also include direct, sustained observation on a long-term basis. Only then will we begin to understand our fellow animals and, in turn, allow them to know us.

Theological traditions, those interpretations of

doctrinal scrawlings, dictate a division between "Other" and "We." Religion has always been a barrier to unity, and the time of faith and prayer is but a blip in the Earth's history. The paradigms of the pious are framed and determined by the dominant population's worldview, including how the universe is put together, be it by God, or gods, or some other deity, or none. Only Nature is eternal.

Those who would follow my path must be innately devoted to the wild and recognize that the animals must allow us in when *they* wish, not the other way around. There, inside their circle, we teach and grow together, as one. We must all trust, all know that we live side by side, not one atop another.

Simultaneously, we must remember that these creatures are forever wild, no matter how welcoming the urban environment. We watch holistically, seeing all as they are throughout the seasons—not as we imagine, not in scientific fragments in peer-reviewed journals. We must see them as complete, living organisms with what we might call personalities—that herein I call "foxinalities"—and know them as we know a friend, as fully as we see and understand ourselves. We see and understand the complete animal, including its character, before casting judgment, before drawing a bead with a rifle scope and hearing that irreparable explosion, that too-late silence.

Magic is real when gray fox females, Tippy and Tense, sit in the same circle with Dark Eyes, who shoved them out into the world and then licked them dry. Blindly, they suckled in silence for the first ten or twelve days before their eyes first saw light as if through fog. At the same time, they heard the wind rustling the leaves. What must it have sounded like when they heard those first sounds, having no previous audible experience beyond the liquid sound of sucking on their mother's nipple?

I sit there on the huge, rotten log in the clearing, surrounded by blackberry thickets and tangles of crisp tree branches, some fallen, dead—nutrients for the land, the Earth—and others reaching for the light. I sit with the ancient ones. I sit and there around me are gray foxes whose lineage genetically runs back at least ten million years—probably longer. Annually, they pass on those ancient genes and the knowledge gained over millennia. This is the crossroads of the landscape, with an ecosystem all its own. No fox, no raccoon, no opossum, no animal at all claims it as its own. All the corridors link together at this point in the big clearing.

The moment I realized that I sat with such ancient beings, I cried. Within that time of tears, I realized how genetically immature we humans are compared to the rest of life about us. I sat there and wondered how much wisdom must be channeled through their community, and how much is lost upon us.

They know of a world so alien to me, too difficult to even begin to comprehend. What would it be like to follow the alpha male, Dark, back into these thickets? What must it look like to follow trails deep into the clearings where nothing but the animals have been, where no human has touched? I imagine being there, following on the hunt.

I came to call her Little One. She was smaller than the rest, yet intense, her motions vibrating, flitting, sudden. She hid from everyone along the levee road while her mate, Creek, sat at the roadside and watched early morning runners as they pounded by on the Black Creek Trail.

She teamed up with Creek, the elder fox, back there along the northern edge of Madera Creek. Her natal den lay hidden somewhere not far past the pump station along the gravel road. It wasn't until she moved the den down to the coyote bush at Palm Tree Corner that I caught a glimpse of her one morning just as dawn wiped night from the land with light, with sun. She peeked from the brush, and then vanished.

Little One had four pups that season. Creek brought them out from the brush and let them play just across the old levee road from where I stood taking pictures and documenting their behavior. They were gray-black balls of four-legged fur. It seemed as though I didn't have to wait very long before their color changed and that rusty red began to define, to take shape as they ran and chased and romped about, beginning to carve out their individuality that, in turn, would cause them to separate and flow into a seasoned life upon dispersal.

Gray foxes live semi-socially at least seven to eight months out of the year, as the pups learn more and gradually extend their range until they no longer return. From what I've seen, females choose which male they want as a helpmate. Mates are largely monogamous, but only to a degree.

I gradually came to realize that nature works in

ways that are not my own, and I needed to accept that. I wanted to pick up Bright Eyes and take her away from the place where she lay beside the road. I wanted to take her all the way out to where her parents, Little One and Creek, lived with their only surviving pup, Dark Face. He was only one of four to make it that far. In clear denial, I wanted to see the pups chasing and playing while the mother and father rested nearby. I came down into the floodplain. There were no foxes in sight. None alive, anyway.

"Probably out in the marsh, hunting." I walked down the old rough levee road.

When I settled into the driver's seat, I told myself that the dead gray fox lying beside the road had to belong somewhere else, had to be something other than the fox I knew. I couldn't look to confirm my fears, couldn't bear another loss.

The next morning, only Dark Face came from the brush. He moved as though burdened, depressed. It took me four days to accept that the body out there along the frontage road was his sister, Bright Eyes, and that Dark Face and Creek and Little One were mourning the death of one of their own.

I mourned with them.

It was only then that I allowed myself to go to the

place where Bright Eyes had died and say a prayer to an energetic young female gray fox who had loved nothing more than to chase her brother through the fallen trees of the floodplain.

All the while, the female that I came to call Helper took care of the food for the den. Helper hunted. She taught the young, but she was barren. She was the sister of Little One, and I witnessed sister helping sister to ensure the next generation. I saw more "humanity" in her actions than most of what I have observed in my own kind.

Dark Face was the only pup I actually saw disperse. One evening, he groomed himself out in the middle of the concrete overflow channel.

That was the last time I ever saw him.

Trust was already embedded in the environment when Squat brought his daughter, the young female that I would later call Mama Bold, out beside him at the edge of the road. I documented their behavior every step of the way. I could never have known that it would be Bold who would one day defy her father, battle him, and win possession of the ancient natal den down there in Fox Hollow—the very den where he had been born.

Bold matured in nine months and had her first litter that following April. Into her third year, Bold and her mate, Gray, raised five pups. Along the levee skirting Madera Creek, Little One and her mate, Creek, had their litter, and they, too, had come to trust me, had found me an oddity that sparked their curiosity. They followed me. I kept talking gibberish at them.

Sometimes I came down low, face to face with one of them. I looked at the world nearer to their level, but that was an utterly impossible world for me to navigate. My human mind had no idea how to begin to perceive it and so I shut it out.

I noticed that Gray spends more time in the natal den than does Mama Bold. It appears as though she only goes to her pups when they need to nurse. I've watched her go, and by the way she carries herself, I can see she's weighed down by reluctance, but she goes nevertheless, swallowed by the brush.

One afternoon as I made my rounds, I came along the old Fox Hollow trail. I knew that Mama Bold was somewhere nearby. She stays close. She was all stretched out under the buckthorn bush, head up, watching me. I took pictures of her under the branches until Gray interrupted us, coming from the direction of the natal den. I watched. He went to her, nudged her lightly, turned, and went back to the den. Mama didn't stir. Less than thirty seconds later, Gray came back into the small clearing as Mama stretched and stood. She looked up the slope at him and vanished into the brush.

When Gray came out and nudged Mama, what was he saying to her? What information did that interaction contain? Maybe he reminded her, "Time to nurse those little ones." Then he went back into the den to check in on them. Maybe. When he found that they were safe, he returned to the clearing to say, "It's okay. Go ahead."

The foxes grew up, had a family, and passed on the family memes—and by "memes" I mean the elements of culture and behavior that are passed down through generations by imitation, not the contemporary, ubiquitous images shared on the internet. One such meme is that, in an urban environment, humans are to be watched but not feared—as long as they aren't aggressive. They handed that meme down through generations, and the receipt feels like this:

1. One adult of the pair is habituated (i.e. accustomed to humans).
2. The adults have pups.

3. The pups clearly see that at least one parent doesn't fear humans.
4. The pups are also habituated.
5. The pups grow into adults.
6. The new adults pass on the same information to their young.

That's how the majority of habituation works. Animals in an urban habitat cannot sustain 100 percent wildness, because they live so near to humans and their bustling clamor. Habituation is inevitable, given time.

Not once over all these years have I ever fed the foxes, yet I've been accused of such on many occasions. My state and city permits forbid feeding. How can someone accuse me of feeding the foxes without a shred of proof? Yet, given some of the foxes' behaviors, I can understand how some might think they approach me, follow me, and permit me nearby because I feed them.

But if we take a look at the larger picture, we find ample evidence of how foxes leave their dens to pass through backyards, through corporate campuses like Facebook and NASA, traveling as far away as the towns of Martinez and Santa Rosa; gray foxes across multiple regions have been shown to be unafraid of people. The red fox population at Las Lagos Golf Course in San Jose is a good example of red foxes acting the same as the gray foxes at the Los Palos Baylands Preserve. They, too, travel through backyards, down old dirt roads, and in the end get used to human traffic crossing their paths.

It is important to remember that these are urban foxes, unlike their next of kin along the wild creeks of the Santa Cruz Mountains, where the foxes seldom engage with or even see people. Behaviorally, the urban foxes display important differences from their wild relatives. They are born into areas where people are coming and going all of the time and thus, they get used to humans from birth.

The gray foxes of the baylands are born all but habituated. They follow because I do not threaten them, and because I talk to them. They hear the rhythm of my voice. I treat them much as you might treat your dog or cat. They seem to understand the names I call them and respond accordingly. We have a bond, and yet they remain mostly wild. Truth be told, they are right to be wary of us.

I wonder about their minds.

How do they think? What happens in those brains of theirs? I wish dearly to get a complex reading of how their minds function. I want to put one of the foxes in an MRI and collect brain scan information, but no one wants to fund that. Wouldn't it be interesting to take a look at a gray fox's brainwave patterns and see what's going on in there? What might we learn?

In the evening, magic happens when I sit on a log in the clearing with seven gray foxes. We are all simultaneously seasoned and callow. A gathering like this is called a "skulk" or an "earth." Gathering is part of their daily routine, to be present, to lay around, to lick lips, to kiss, to scratch, to nap. They allow me to be there nearby. Magic rustles the bushes, the trees, the blackberry thicket when they are about. The vibrations in our moment shift. There seems to be another way to see, a kind of mystery humming, silently singing, in such sacred moments. Here I sit in the midst of these predators, these gray foxes, these bush dogs, the most ancient of all canines, and they have come to this place. I am not quite one of them, but not quite separate. Why do we gather? Why do we sit here together? Why have they permitted me?

Tippy, over by the walnut tree, had lain down several minutes ago and closed her eyes. She needed a nap. Her sister, Tense, sits watching by the trail. The alpha male of the region, Dark, sits in a proper gray fox pose with his front paws side by side, like a sphinx. He always smiles, even when he's vicious. He radiates a cool aloofness—his foxinality—not found in any of the other foxes that I've ever known.

Dark is involved with himself. That feels dangerous, but it's the way he governs this region.

All of the foxes are there and when they speak, they do so with gestures, not a sound as we do. The yearling female sisters, Tippy and Tense, come to greet him with submissive, down-on-the-ground, bushy-tail swishing and snakelike body movements, ears flat back on their heads, and rise beneath Dark's chin with a lick. Wet.

I can only watch and report, but a part of me wishes to be one of them, to crawl across the grass to nudge and nip and dip my head in this ancient ceremony.

Dark taught me everything about what it means to be an alpha gray fox. As head male of the region, he had a duty to all of the females that when they approached estrus, he would remove any lingering males. Only Dark's genes were robust enough to pass along; all other males were simply inferior. Dark culled the gathering. He chased the odd males away, except for his son, Blue, who had dispersed and returned to find himself tolerated but unwelcome.

Tense is emotionally powerful. When she meets another that she does not like, she takes on her deceptive submissive greeting: low to the ground, looking up in a semblance of supplication, but then baring her teeth. It is the devil's grimace. Submission becomes aggression. A rip in her face and all forty-two teeth exposed send shivers. Issuing a rocky growl from her lowered position, she threatens.

Tense gets her teeth in Dark's face. She wants to become the alpha female of the group, but her competition is the alpha male and his counterpart, Dark Eyes. She doesn't know how to handle them, but she still tries. Dark has ruled this home range for the past four years. Tippy wanted it too, but she found that following a young male was clearly the best choice. Tippy vanished from the home range as she trotted along with that interesting male. She dispersed. The raccoons, the opossums, the woodrats, and all other wildlife in the region kept their distance.

One year further, Tense was chased out. The alpha male kept thumping at her, kept shoving her, chasing her every time he saw her. He kept abusing her until she could no longer hold her ground. Why Dark treated Tense like that, I don't know. He chased the males and that was his duty, but to chase a female like that raised so many questions. She gave up and went over to the north side of Madera Creek. There she teamed with the adult female, Helper, the adult male, Brownie, and others on the north side. They accepted her with a sense of well-being throughout their home range. Brownie and Helper kept the peace. Tense's sister, Tippy, followed soon thereafter.

Did Tippy relocate, chasing that alluring male, because she missed her sister, or did she grow tired of Dark's aggressive rule? No one seemed safe from Dark, who would eventually injure himself just to rid the home range of his own son.

Blue had been tagged by Dr. Brian Hoper from the USDA Office by the Don Edwards San Francisco Bay National Wildlife Refuge. Someone had hired him to tag that gray fox over near Shoreline Lake. No one seemed to know why—not even Dr. Hoper. Blue was old already when he'd come into the Madera Creek jungle. Dr. Hoper told me that his records showed he had tagged the fox called "Blue" two years before I met him. It took Blue two years to cross over from Shoreline into the Madera Creek thicket—no great feat where foxes are concerned. That's like moving next door.

Back in the spring when he first arrived, Blue and

his fellow mammals were being eaten alive by pests—constant scratching. Tense would walk down the channel, stopping every ten feet to pump her hind leg at a cluster of fleas or red mites. Ticks, chiggers, and all manner of vexation inhabit that warm channel of the ear that runs deep into the gray fox. This discomfort is an annual event in their lives. They are born scratching.

A ritual practice among the gray fox (and maybe other canids, too) is ear cleaning. I have only seen it in the gray fox, but I am certain that it extends to others. Since they cannot clean their own ears themselves, they often trade off. If they talked, they might say, "Hey, I'll clean your ears and you clean mine."

"Agreed, because those bugs in the ears taste so, so good. Be glad to clean yours as long as you clean mine."

The process isn't always done out of equality. Once, Dark came in on the Ivy Trail, back into the clearing. His mate, Cute, was already there. She went to him as he came in and rubbed her head alongside his, suggesting that he clean her ears. Dark remained sitting in his royal pose, so Cute went to him and cleaned his ears, lapping up the vermin that was filled with essential nutrients and protein. There was no reciprocation.

But the ears are most critical because, if their ears are clogged with vermin, then the foxes cannot hear as well as they must—and hearing is their primary sense. It is the auditory sense that allows them to triangulate on prey deep in the grass, to fixate, to listen before bounding in a gracefully arching leap to come back up with a rodent clenched between their jaws. They may have an auditory "map" in their mind. If their ears remained filled with vermin, they would be severely handicapped, but also likely hungry. Their ears are paramount. They use them like we use our eyes. When watching, their ears are always in motion, scanning, or one ear is locked in on a sound over to the left and back, while the other ear is listening for danger off to the right. Their power to hear is responsible for their survival all these past ten million years that they've been on this planet, living through its many upheavals, its ice ages, its warming episodes, through feast and famine.

I suggest that there are multi-processors of information streaming into their psyches all the time, not unlike ourselves. However, our sensory evolution has led us down different avenues. Sight is our most often used sense. We see the world around us and that translates into feeling. From there, most of our senses fall away. For the gray fox, however, the story appears much differently. In their minds, as they go about their time on Earth, they create an auditory universe of their region and that audioscape is enhanced through smell. The complex mix of odors and sound create a brainwave universe of understanding for them. They understand trees as auditory and olfactory elements translated into paths, along which there are connections to food, water, safety, and rest. They understand the hunt by listening and smelling the field mouse, thus forming auditory and olfactory images in their mind, formulating strategies in real time as the mouse scurries about in the dry weeds. Knowing that hunger—that drive to fulfill a biological need—creates action. That's how they think, or so I suppose.

Before Dark forced Blue out of the region, Blue would compulsively clean and re-clean all of the females' ears. This was his only way of being around the females, even Cute, Dark's mate. Blue had a gently aggressive side that he showed when he cleaned ears. He'd often plant his paw on the side of a female's head to hold her down so that he could lap up the vermin without her wriggling around due to his invasive tongue. Once he'd feasted, he'd smack his lips.

Every year when Dark saw Blue in late February or early March, the chase was on. Two gray foxes crashed through the brush, breaking low branches as they flew. I could follow them by that sound alone.

On most occasions, they crashed downstream along Madera Creek. They did not have time to follow a trail. They blindly tore through the thicket until Dark either decided he'd had enough or had injured his hind leg enough to limp. Anger grew between them until, like Tense, it was far too much for Blue to handle. He just wanted to be left alone, but to Dark he was a threat. Blue broke. He left his home and went over to the north side of the creek. He joined the others.

There in that new mix of foxes is where Blue's foxinality changed. On the south side of the creek, he was active, moved with force, felt grounded, and had an identity; on the north, he went from an upbeat, extroverted gray fox to a timid, withdrawn coward. I could hardly recognize him. It was as though Blue had turned himself inside out.

He hung out right along there on Palm Tree Corner. When Tippy came along he felt drawn toward her, but she wasn't ready to feel anything for a male, so he followed her instead. Blue no longer cleaned ears, except on rare occasions. He distanced himself, but Tippy seemed special. He followed his curiosity.

I wanted to install some new trail cameras down along the creek, to see what the foxes were doing at night, as well as during the day when I went "foxing." These gray foxes are crepuscular; they don't just head out into the night like the raccoons and skunks, but they are also active off and on during the day. Trail cameras were the obvious answer to some aspects of ignorance, and at that time, I was unbearably ignorant.

That first year when I installed the cameras, I hit jackpot upon jackpot—a veritable gold mine of video images. Over on the Landings Road biofilter, under the canopy one night, the five pups came dashing into the clearing and there they played the whole night through. They showed me how they wrestled, how to efficiently cinch jaws to neck, bite, and shake to death, how to be agile and climb high into a tree with ease and speed. That gave me confidence and taught me how to best use the trail cameras.

I didn't move. I peered through the willows. Below was exactly what I sought: that enticing clearing. I carefully planted my boots on the slope, one step at a time, one boot on the steep slope, until I hit bottom. I grasped the trunk of a black walnut just beside me and relaxed. Across the way and off to my left, a two-and-a-half-foot thick tree trunk lay. Grasses grew, along with mustard weed and some exotic plant that the English call "lords-and-ladies." Its flower lasts no more than a day. I saw plants that were wonderfully green and robust with life but their names were nowhere to be found. Silence enveloped everything. I stepped through, maybe ten feet or so. Near my feet an animal trail peeled off to my right. I went over to the big rotten log. I wanted to sit down, but as I looked, as I pulled away loose bark, ants boiled forth. I stood there, frustrated because my quest for comfort had been defeated by ants. I turned back to the big clearing and stood there surveying it all.

Frustration washed away as I saw the entries: five holes, five connecting corridors, branched from the place where I stood. The thicket trail seemed to have the most traffic. The big clearing took on a life of its own. The foxes already knew it well. They came from the thickets when I emerged, step by careful step, into the clearing. They gathered.

At night with at least two, sometimes three trail cameras recording the corridors, those highways that passed through were busy with traffic: opossums sniffing, finding, digging, and cleaning themselves, washing their faces; a mama raccoon followed by her young ones who ran and played and wrestled and cavorted with each other until Mom was so far ahead, they had to take a break from the fun and catch up. Sometimes, when they were truly far behind, they sent out a soft rolling cry, calling for Mama.

I had two cameras under the canopy to document every detail of the gray foxes' schooling. I call it "school" because they were learning life's lessons and honing their own survival techniques. What human playground is any different? There in the big clearing, pups aged, learned, and dispersed. Midget was one. He belonged to Cute and the alpha male, Dark. He was the only pup at the time, which made it hard for him to develop because he didn't have siblings to learn with, and so he befriended one of the young foxes from over at Dark Eyes' den: Tippy. They hunted together, they wrestled together, and they enjoyed each other's company, lying side by side there near Thicket Trail. Sometimes, in a flash, they fought with sharp yips of pain and screams meant to frighten.

When it came to anything related to food, friendship dissolved rather quickly. Greed dominated. Tippy sent that low rolling growl as a warning not to come any closer. Midget took a step. Tippy growled and bore her teeth. The field mouse lay off to the side near Midget's left. He growled the same kind of growl and hiked his tail, showing that he was ready to fight.

In less than an instant, they collided with shrieks and yips and cries that only a gray fox can make. Within five seconds, the battle was over. Tippy limped away as Midget bit into the field mouse's skin, enjoying its juices, savoring his tasty meal.

FIVE

THE MIRROR

I asked myself questions about these gray foxes, such as: "What will their reaction be when they see what's in the mirror? Will they recognize their reflections?"

To find out, I looked to the foxes at the water purification plant. Along the edge of the Landings out to the Los Palos Baylands Preserve, one fox family commanded the hill right beside the intersection of Landings Road and Harbor Way. That family was Ears and his pups. I never saw an adult female anywhere with them. She may have been killed, necessitating that Ears raise the litter as a single dad. He boasted four pups. One of them I dubbed "Fearless." She was what I considered a legendary alpha female, because from the time that she was a tiny ball of dark gray fur, she showed no fear whatsoever. She didn't know what it meant to be skittish, although she would never let me get closer than maybe fifteen feet away. She became the alpha female cub of the family.

At night, the pups dashed and chased and climbed trees before the video trail cameras, those motion-sensitive cameras, and they caught Ears and his family down beneath the canopy where fox school took place. The pups cavorted about, swarming into the trees, wrestling, picking up sticks and tossing them around. Under the canopy, the pups particularly liked to fling a discarded pair of ragged men's shorts, learning yet having fun. Their cavorting made sense; it was practice for when they became adults.

I devised an idea: set up a mirror and observe their reactions when they see their reflections. Would they see an alien, something neutral and harmless, a frightening figure, a welcome image...a friend?

Late one afternoon, on Tuesday, July 24, 2012, at 5:34 a.m., I staked out a mirror on the sand filter. The mirror: 18 by 24 inches, enough for them to see themselves life-sized. I knew the foxes would come when they saw me there, so I stood patiently, waiting for something to happen. The adult male of the den came from the hole in the brush, the hole right beside the dead eucalyptus branch that had fallen last summer.

Ears saw the mirror. Curiosity drew him. When he looked into the mirror, he must have been stunned by an explosion of adrenaline, for he dashed across the sand some twenty yards or so. He stopped. He turned. Slowly, he read the sound of the environment as he knowingly walked back toward the mirror. I often wonder what he thought, why he decided to come back and take a second look. After all, the mirror had stunned him. He should have remained frightened. Fear should have prevented him, but it didn't. How did he overcome it? He returned to the mirror. He looked, started, stopped, gathered himself, and came in even closer. I would have given anything to be inside his mind at that moment. What was he thinking? What stirred?

Meanwhile, his four nearly grown adult pups came out to see what Dad was up to. They, too, looked in the mirror. When they saw their images, some dashed off across the sand or back down under the canopy. Fearless simply stood and looked. She pawed at the image, sniffed it, and walked around to the backside of the mirror to try to find that other fox. What kind of thinking does that take? Within just those few moments, their universe of knowledge bore down on trying to understand something that was 100 percent foreign.

It must have been terribly confusing, terribly troubling.

That next evening, I installed the mirror down beneath the canopy, where the fox family came together. I set my two Bushnell trail cameras on the mirror. That first night there was an assault by the gray foxes on that intrusion into their lives—the cameras and that reflection. That early trail camera emitted a dim infrared glow, revealing animals shaken by this disturbance. The young foxes physically attacked the camera and the mirror. They leaned on it, sometimes two or three at a time, until they pushed the mirror down. When it lay flat on the ground, they walked around on its surface, looking at themselves below. Were they intrigued by what they saw? I don't know, but soon thereafter the family all left, and by that I mean they vanished from the region. Why did they leave? Did the mirror cause it, or was there another catalyst?

I saw tension rising in the gray fox community. They remained clear of the camera and the mirror. Apparently they did not like that mirror anywhere nearby. Once I finally saw what was happening with them, the unease it caused, I removed it.

The next morning, the foxes were far more relaxed.

Did they know it was gone for good? If so, how? The experiment raises a thousand ethical issues, brief as it was. Should I have introduced the mirror in the first place, given that it set this gray fox family on edge? Surely I shouldn't persist in something that causes discomfort to wild animals, but was it wrong to test their curiosity? How could it be?

SIX

POISONING THE LAND

Along the creek, there are foxes living and cycling through. Generations have passed. This is a pinch point in the land. The terrain is such that it channels all of the wild animals into a funnellike jungle of thickets and clearings along the way. Moving upstream, moving in a northerly direction on the banks of the Madera Creek, the brush pushes into the levee road, breaks out into a brief floodplain, and then slides up to the bridge and the insanity of Highway 101 North. Cars, trucks, and motorcycles all shove themselves through time on that freeway north and south, east and west. Along the creek, all that takes a backseat.

The gray foxes that live in the thickets don't quite

match classic descriptions. They are classified as solitary animals living in a quasi-communal setting due to the human-made pressures on their environment. The urban gray fox is not a solitary predator. Often a mated pair will hunt together, and of course raise their young together. So, within the family system, rather than being solitary, they are social. But when their natal den territory is severely compromised because the region is overpopulated, some of the adults will go so far as to help raise pups that are not their own. It becomes a communal gathering. Some foxes get along well with others. Is that just their nature? Sometimes they are social and sometimes they treat each other as invisible. Then there are those who come into the circle of foxes and bare their teeth.

These were the gray foxes: the sisters with no names, and Gray and Bold, who lived a double tragedy of having their pups decimated, killed, murdered by a poisoned rat from over in the technology center next door. Moreover, all because the people who are in charge of those buildings know not what they do except to mind-swallow the sales pitch: anticoagulant poison bait boxes, sometimes three or four along the outer walls of a high-tech building, just next door to the foxes' home range. Management gave a command to keep the rat population under control, so in came the anticoagulant that creates suffering here in the wildlife community and in others across the nation. It's so thoughtless. An easy, non-murderous solution? Attract barn owls into the region with owl boxes. They will normally devour upward of six rodents per night. When they have babies, they will kill twelve to fifteen every night of the week.

I wonder, did Mama Bold watch her five pups, one by one, take their last breath as the poison consumed them?

I aged. I was no longer that fourteen-year-old who had trapped gray foxes. Time altered my life like stew in a slow cooker. At the same time, I kept tripping over myself, burning myself due to emotional clumsiness. The foxes became but a flicker and even dissolved down deep beyond memory. A cultural abyss sucked me in. Gray foxes no longer lived in my consciousness. The foxes disappeared.

My future lay ahead; things I thought I wanted to do and somehow make a living doing became a roughed-out image of who I might become. I was struggling and stumbling along, trying to put my life together, failing to realize that all I had to do was relax and look my reality in the eye. I had no idea what pattern life might take. I knew that I had a university degree to attain—a PhD, I thought—a family, and children growing up all around me, but most of that tumbled into oblivion in the hurricane spiral of a divorce.

In the unnecessary fire and fury of severing that deep bond, I lost touch. I crumbled. I realized that I was not an especially good father. I didn't know how to relate to my children—not until they could think, at least. Something had slipped between us. That was what I felt; the rest was far from consciousness.

I was in my late sixties after I quit teaching high school English literature and writing courses so I could help to grow a high-tech start-up as the president of a dysfunctional and underfunded company in Silicon Valley. Two start-ups later and without much going on in my life, the gray foxes called for a second time. Having heard, my mind and my core chimed in with a fascination about these canines that act like felines, these brash bush dogs that live in the brush, that are for most people out of sight, out of mind. Circumstance dictated that I follow.

When people see a fox, they fall in love and want it

as a pet. For me, I needed to understand those foxes as I watched and heard their foxsinalities rub one against the other.

They blossomed into an addiction once more. The more I came to know, the more intense my need to know took hold.

I accepted their challenge. I knew that if I was to understand these urban gray foxes' behavior, somehow they must allow me, must trust me inside their community. They needed to accept me. Instinctively, I began interacting with them, talking to them, showing that I was noisy while they were nearly silent, much the same as I did with my dog Stuffy back when I was a kid. By talking to—and often with—these foxes, their trust in my actions, my scent, and my voice grew more and more familiar. I allowed them to trust me. I became a part of their environment, another being. In many ways, in order to gain entrance to their culture, I had to pass some kind of a test. The problem was that I was never aware of the tests until they had already passed. For instance, I was to remain careful while in their presence, not move too quickly, respect their space, and do everything slowly, even to the point where if I wanted to take a picture of one, I had to take great pains to ease the camera to my face. Once they grew comfortable with me in their midst, then they trusted me.

In somehow passing these stages, I saw that the local foxes began to relax until they were no longer skittish unless I tried to come too close. They trusted me. Often (but not always), if a stranger entered, they hit the brush running. I, on the other hand, was allowed to see the inside of their culture. They were semi-habituated: they to me, I to them. It was then that I knew I could certainly come to understand them, given time.

Again, they taught me. After three divorces, I was through with that stage of my life. Now I ran truly free, driven by curiosity because the basic systems in my life like shelter, clothing, food, and the ability to pay my bills, all flowed. Life, for once in my time here on Earth, became free from inflexible bonds.

SEVEN

LETTERS AND NUMBERS

I have been criticized for giving each of these foxes common names, like Gray, like Mama Bold, like Pale, and like Blackie, who was visually and behaviorally distinct from any other gray fox I came to know. He wore a raccoon mask with tapered, slick edges back along his cheeks, so prominent that I could instantly spot him in any of the black-and-white night video footage from my trail cameras. It was as if he'd stood up and announced, "My name is Blackie." There was no other name by which I could know him.

The dictum is that, in order to remain scientifically objective, I must identify them with letters and numbers. You know, so everyone who knows the code gets the same information. "GF-23" means that, as long as that fox is tagged with such a number, it will remain an objective animal. Pure science, yes?

Not true when one allows themselves to feel these individual foxinalities. GF-23 can just as easily become known and loved. Look down in southern California, in Griffith Park where puma P-22 roams. He is the most loved and famous puma in the world because people have learned to see his emotional side. He is not an object, and yet largely he is within science's bounds. P-22 and OR-7, the gray wolf, are individuals, especially when you come to know them and they to know you.

I stand outside the dominant scientific paradigm, and I am *comfortable* here.

I was so fortunate. I have been so fortunate. From my unique vantage, I can attempt to interpret the meaning of the foxes' behavior. I can never be sure, yet find it compelling to see just what I might eke from this darkness, from this ignorance. Along the way, they drive me to understand by unfolding themselves before me.

EIGHT

ON LOSING PUPS

It turned out that the ancient natal den where Bold had been born, the den she'd fought to win from her father, was the place where death would claim two of her litters. Twice she lost the little ones there, but when she later gave birth to a litter in the new den some distance away, the litter survived. Why? I wonder if she could discern that the den in Fox Hollow kills, if every time she'd had pups over there along the road, heartbreak struck. Did she experience heartache, or did I project that upon her? She may have thought, *I was born there and deposed my father for its ownership, but that den kills my young. The den up on the hill under that alkali saltbush seems to let our babies live, so that is where I shall keep my babies safe.*

NINE

WHAT DOES IT MEAN TO BE WILD?

What does it mean to be wild? What is the difference between being tame and being wild, civilized and uncivilized? We can go through so many millions of words on the internet that seemingly address the definition, the explanation, of being wild, but the glut of supposed information and opinion makes it all meaningless. Instead, from Merriam-Webster: "living in a state of nature and not ordinarily tame or domesticated." "Wild" then becomes anything that lives beyond our civilized morality, anything that does not live by the same rulebook as do we modern human beings.

This word and all of its connotations have an

important historical background flowing back to the age when we first broke our bond with the wilderness, divorced ourselves from our birthplace, dispersed, and evolved along a pathway that diverged from the wild. Most likely it happened when we decided that a goat herd and a parcel of land weren't quite enough to allow a woman, a man, and too many children to thrive. I say "too many children" because the dictum should not be "Multiply and replenish the Earth," but "Rear only those you can adequately provide for." Settlement altered the ways we human beings relate or don't relate with others, and even the ways we relate to ourselves.

What are the ramifications of civilization? In the common parlance, if something is wild, it can be killed for sport, for food, or simply for being a nuisance. The modern view of animals like wolves, like coyotes, and like other apex predators, is that we have demonized them so that killers—predators with high-powered rifles, scopes, and a bucketful of tricks—can come along and wipe out a whole species. Might it be that we see, for instance, the wolf or the grizzly bear as reflections of ourselves? Our fear of ourselves has, in this regard, become lethal. We kill the wolf to feel safer, unaware if they are truly a threat or not, unaware of the deeper impact of such actions.

The stories run deep about predators. Before we made that leap into what we now call consciousness, reasoning, living ethically, we were their prey—they preyed upon us. They broke from the dark and devoured our own, chewed on us in the night. What fear that must have cast about. What ideas arose to defend oneself? In the end, that fear ensured ignorance because it became locked within. Many held that the predator loomed dark and dangerous. Their egos were infected by fear and that infection caused the first etching depicting separation.

We forget that we, too, are animals—mammals, alive on planet Earth, living side by side with other aware beings that we are remiss in ignoring. All sentient beings on this planet are thoughtful, warm, and tender, furiously and sometimes painfully battling for every moment of life. Each one is worth fighting for.

How did we ever grow into thinking otherwise? How did we place this barrier between us? Someone lied to us, making claims such as "Human beings are the only animal that…" but if we'd look again, we'd find such claims invalid. We defined ourselves as the world's only toolmakers, with larger brains, opposable thumbs, and so on. We are not the only creatures to possess such traits, and it is pure arrogance to think otherwise.

The will to survive lives at the taproot of who we are in relation to all else that populates the Earth. Long before the appearance of spears and other such weapons, human beings were cunning enough to stay clear of those animals that could claim them for dinner. These ancestors died alongside the others, on the banks of a slow-flowing river, in the burning desert sand, in a marsh, and in the forests deep, all for the same reasons.

Through the twists and turns of millenia, human beings have alienated themselves from the wild in order to survive because human survival depended upon being able to hide, or barring that, to defend, to retaliate. We made ourselves the planet's most aggressive and invasive animal, and, in the process, we destroyed huge swaths of land across vast regions far and wide. In the places where the young should have been born and licked dry by she who gave birth, we instead killed our perceived enemies, wore their skins, and spread our fear-laden stories about the dangers of the wild.

Bob Rummel is an authentic mountain man—and my good friend—who lives back in the wilderness of the Badger-Two Medicine, which is sacred to the Blackfeet Nation as the birthplace of the tribe. I was introduced to Bob because I wanted to know how global warming was affecting the high mountain regions where he lived, just outside of Glacier National Park. We exchanged emails, and in one he invited me to come up to the mountain and spend time with the elk, the bears, and all manner of wild critters—an offer I couldn't refuse. Once I'd arrived and it was time for dinner, Bob would set the firepit ablaze and wait until the coals were just right, and then he'd lay on a couple of steaks or some other form of meat. We kept the fire going long into the night. On one such occasion in 2017, I asked him, "Hey, from your viewpoint, what does it mean to be wild?"

He leaned over and stoked the fire with a hooked stick. I watched those cinders sparkle toward the aspen branches above. The fire flared. Coyotes called from one mountainside to another across the way, on up the glacier-gouged valley. Bob kind of hunkered down and smiled with that sidelong smile of his.

"Wild, huh?" He looked up. "In my opinion, 'wild' is anything that doesn't adhere to our human social and legal structures. If wild, it is not constrained by Our Law, Our Ethics, or Our Morality. None of that. It lives beyond those gates and fences that we erect and that separate us from them."

As I often did, I found myself agreeing with his words.

In the deep winter, snow comes up to the windowsills.

It's -18 degrees Fahrenheit out there. The wolves are out on Dog Gun Lake in Montana, on the Rocky Mountain Front. There, Bob watches them, enthralled, amazed with a rush of happiness that courses through his mind and body. I know it well. He phones me and says, "Hey, there are five wolves out there on the ice right now. They look like they are on the move." How special that phone call was; how extraordinary to hear a like-minded soul.

Since around the age of thirteen, I have been taken on a journey by the gray fox, the likes of which I could never have imagined, not in my most intense dreams. The more I probed the behavior of the gray fox, the more intricate it all became. I can't speak in generalities because when I do, I am telling lies. I lead you astray of the truth, the kind that society sanctions for its academic researchers. I might add that we also need to examine the paradigm that science is working within, for it defines our worldview.

I've observed only the tiniest drop of wildlife in my time. I've seen a group that felt the impact when too many creatures lived too closely together, when the virus consumed a large swath of one region. The fall of one ushers in the imbalance and dissolution of the whole. None stand alone.

If your life brings you near the wild, you may come to

know enough, to remember that there are plants to eat and enjoy along the way. Sometimes I try to introduce others to the tastes of the peppery wild radish and the breath-cleansing fennel seeds that fill my mouth with licorice.

TEN

THE TRAGEDY OF BRIGHT EYES

The family consisted of one helper female, two pups, Little One (the mother), and Creek (the father). The male pup I called Dark Face and the female I called Bright Eyes, because she always looked so alive. Her life danced in her eyes. They were brother and sister, these pups. They sometimes broke from their play and came up on the road toward me, especially when I called them. I knew them face to face.

In the dawn's light, at a time when I could just begin to make out shapes against the shadows, the world of willows, of great palm trees, of blackberry thickets, of the small floodplain, and of the creek began to take form, and a new day was born. The gray foxes came back home to settle in and begin their first nap time: our morning. Down in the channel, they napped into the early afternoon. That shift in light and dark is their clock. We hardly pay attention.

As the pups began to explore the world about them, every step outside their home was a unique guess. From that deep thicket, they slid through the brush and out onto the concrete channel. Across the creek and parallel to the channel, the Black Creek Trail skirted the creek to the pump station.

Bright Eyes and Dark Face rejoiced with one another. They ran and chased, their hearts pumping wildly as they climbed trees, almost dancing through the region with each other—at one with each other—wrestling, thus teaching the other how to survive. Back and forth, they grew together as they aged.

Out in the Madera Creek floodplain, there stood a stand of dead willow trees, mere sticks for the pups to prance through as I stood watching. Their play was amazing. Dark Face and Bright Eyes dashed about through those naked branches, chasing and playing with each exhilarating moment. She'd launch after him and then he'd turn and rush toward her. Often they'd stop, face one another, bobbing their heads, moving from side to side as though trying to get past the other, and then the chase was on. She ran at precisely the exact moment. He chased her up into the tree, then back down only for them to wrestle with each other below that dead willow. They taught each other how to be good gray foxes. They learned to be survivors.

Away from the floodplain, out there on the asphalt road, Dark Face chased his sister long into the night. Heedlessly lost in their enjoyment, they dashed out onto the road and were met by a car. Bright Eyes didn't see it; the headlights blinded her. She whipped around, her head glancing off the bumper, and flew to the roadside where she slid to a grinding halt. She lay quivering, every thought shaking, her body burning, her mind spinning, unable to understand as her body shut down. She lay beside that frontage road, in fear, for as long as it took to die.

The driver of that car probably didn't see either of the pups in time. There was no "Fox Crossing" sign to warn travelers. They may have shrugged it off as an odd bump in the road and driven on, their attention on getting home before it could get any later. They didn't consider that a member of the gray fox family had been seriously injured, that they'd just brought sorrow to the foxes living along the banks of Madera Creek. Might the pup have lived if they'd stopped to help?

Or maybe the driver shrugged, "Who cares? Just some fuckin' stupid animal," and punched the accelerator. I will never know, but neither will the family.

Dark Face had no idea why his sister had suddenly stopped playing. He lay in the dry grass beside the road, licking his left paw to soothe a strange twinge of pain. *Something* had happened, he sensed; he knew that much. Beside that dark road, he saw his playmate lying so incredibly still. She didn't want to chase anymore. She didn't move. Instinctively, he knew that she was no longer what she had been. A heaviness overwhelmed him in a shuddering wave. Some of that feeling lodged. It stuck inside and weighed him down with deep, suffocating woe.

Sometime later, he rose, intending to trot back down the levee, to go back to where his mother might be sleeping, or maybe find his father, Creek, over near the big eucalyptus tree. Dark Face needed them just then.

His left paw came down. Fire flashed throughout, and he stumbled. After several shaking attempts, he managed a slow and painful limp. The fire in his paw grew too much and he stopped beside the road. He still didn't understand, but he pressed on toward the den, toward his family.

What he found instead was that they were out for

the night, filling their bellies, and no one was home to comfort him. He was utterly alone, save for the two new aches in his tiny body.

Before dawn that morning I drove along the frontage road to my parking spot along Madera Creek. I saw something in the dark, over at the roadside, that looked big. Raccoon, maybe. I shook it off. I knew that it was an animal, but that was all.

I pulled into the parking area, swung around beside the coyote bush, and stopped. I flicked on the light inside the car and filled in the heading for my daily log—time, temp, calm, breezy, windy, clear, partially overcast, overcast, dark, or dawn. That formed the foundation for the Gray Fox Log of more than a million words, a small portion of which you're reading now.

I stood beside the car for several minutes, thinking about that dead animal down the road, less than a quarter mile away. I decided that I wouldn't go down to see what it was. It would take too much time to walk all the way down there and back, and besides, I had data to collect. But something inside drew me down the path beside the road. As I neared the gate, that same something tugged at me to go see about that animal. After all, I needed to contact the rangers to have them come remove it. I'd have to know eventually.

My initial fear was confirmed as I neared the carcass: there lay a gray fox. It was off the road, but not far enough to avoid being hit multiple times. This was desecration. I gently pulled its leg aside and saw that it was female. There was very little blood on the side of her head. I hefted her over, further from the roadside; I didn't want to see another car hit her. She weighed more than I thought she might while in life. I told myself, *This isn't one of the foxes I know, and that's for sure,* and I justified myself with so many reasons why that dead gray fox could never be Bright Eyes or Dark Face or any of the other foxes that lived along the creek.

Bright Eyes was down the road, off in the weeds of the floodplain, playing with Dark Face, her brother. I wanted my disbelief to overcome reality. I shook my head. Because of the body's condition, I couldn't tell which fox family it came from. I looked out across the Saltwater Wetlands. I stood convinced that it came from over by the ITT Facility out there in the marsh.

I didn't recognize her.

Looking back, I realize that I was in denial. Three

days later, Dark Face, her brother, showed up alone. He came from the brush, walking slowly and heavily as he looked at me. He slogged along as he crossed the road and gazed out across that broad marsh. The way he carried himself, I could see depression weighed him down. I had been there in my life and I knew what Dark Face felt and why.

She's dead.

As much as Dark Face had to accept her death, so, too, did I.

ELEVEN

FOR THE FUTURE

Over these twelve years, I have visually documented the behaviors of the urban gray fox in the baylands along the San Francisco Bay. These foxes allow people to approach. They, overall, are not afraid of human beings. That gives me an advantage as I document their behavior because I see their behavior played out right before my eyes. I see the young ones, the pups, nurse from Mama Bold down there in Fox Hollow. I see them spit and growl at one another on occasion. If these were not urban gray foxes, I would be very hard pressed to see them at all.

Aside from twice-daily visual observations of three particular urban gray fox "families," there is an array of fourteen digital trail cameras documenting wildlife events at critical junctures along the edge of the San Francisco Bay and Highway 101.

In the early morning, I head out to collect the SD Cards from the trail cameras. I leave the cameras off and tucked away in the brush during the day because, with an increase in wind during daylight hours, I often caught a branch moving nearby whereas at night that very branch didn't twitch. I stopped setting them up 24/7. Instead, I drove out in the afternoon/evening and set up my cameras for the night. These cameras have contributed greatly and directly to these particular findings. They give a wider perspective on the gray fox because we see them during the night, carrying a cottontail, a woodrat, or some other delicacy. Those trail cameras feed the theme of this report.

I don't see any equations in this whatsoever. There are no numbers to back me up. These foxes are not mathematically defined, sanitized, objectified skeletons. They are not specimens without will. It's been suggested that I remove my own interpretations from the matter. Instead, I must rely on interpretation to understand, because no one has applied such protocol as mine: in the field twice a day, taking notes, watching a hunt or the tickle of a belly in progress, out there in the brushy regions of fox-land. I'm there to see Mama Bold's nipples show; her milk is coming in. She will have pups within the week, or so I predict. I write it all down on my Post-its and tuck them away for…what? For the future, when someone with more acuity and resources can sift through the tangle and take another step toward understanding them? I wonder if that will happen, but I know I have to do this now. I can't ignore the need.

TWELVE

PARENTING

I continued as though no limits had been set upon me: no fees, no red tape, no imaginary boundaries or distance restrictions. I would do whatever it took to be there when they fought, when they met one another with a fox-kiss, when they were consumed with tail-swishing joy, or were drugged with depression, or were withdrawn due to sickness. I would see them learn from their parents, mature, and become parents themselves, year after year.

Gray fox parenting covers a variety of styles. Cute and Dark could hardly tolerate parenting, whereas Gray and Mama Bold had it all worked out. Some might say Cute and Dark were not fit to be parents, but out there in the wild, there is no way to stop the process once it's begun. The gray fox female may have up to seven pups each year, according to the literature, but Cute had only a single pup two years in a row.

I sat there on the stump I'd hauled into the edge of the Rose Den and watched when Cute's pup, the one I later called Big Guy, wanted to nurse. A strong majority of the time, Cute turned away from him, causing her baby to whimper. When she sat back to scratch, I saw the blood on her nipples—only two, whereas normal gray fox females have six. She seemed to be in serious pain.

Midget was Cute and Dark's firstborn, and the following year came Big Guy. Both of those pups needed to grow up by watching others in the vicinity, but they had no other siblings or peers.

Cute and Dark had very little idea how to be parents to their newborns. They twitched nervously within the parent-shell. They shunned their roles as parents, apparently because they didn't know how to do better. They often left their little pups alone, especially Big Guy, forcing them to fend for themselves. I saw it often on my trail cameras out there at the Rose Den.

Big Guy played with moths and other things in the night, all caught on SD card files. His father, whose duty it was to fetch a rodent or two during that time, seldom showed up. The few times when he did, he played with his pups, first Midget then Big Guy later on. It was clearly a treat for them, because he was so seldom around. Tense was there, but only in the background, only as a watcher. She was always on the edge, always ready to bare her teeth and hiss defiance to any fox stepping a paw out of line.

Cute left her pups at the natal den for three or more days at a time. Watching this, I realized she and Dark were absentee parents, never home, so both brothers had to grow up without them, one year apart. Midget matured and left the den. That next season, Big Guy had to develop his own style of living, for he had no role models, no siblings on hand to hone his skills. Big Guy was on his own, and it showed in his every action.

He failed to disperse and his father, Dark, failed to chase him away as he had with Blue, so he was there when Cute came into estrus. He mounted his mother. She screamed at him, screamed a long, piercing "Noooooo!" and bared her teeth in defiance, but he didn't stop. As he mated with her, she quavered, swung her head around, and looked up at him hunched over and rutting like millions of other gray foxes had done. When he finished, Cute ran from her son, crashing through the brush with him directly behind. I followed and tried to capture this unprecedented event.

I filmed. I lost them. I broke out along the edge of the brush, and far down the channel there was a fox.

From that distance, a good two hundred yards, I couldn't tell which fox it was, so I walked in that direction, still stunned by what I had just seen and caught on my camera's video. I walked out onto the concrete channel and trotted toward the fox. It was Dark, the alpha male. Cute's mate.

When I reached him, Dark turned toward the patch of tall weeds nearby. He walked over about ten feet or so, one ear trained directly ahead, one twitching a bit off to his right, listening. He crouched and then, in an arching vault, plunged into the weeds. He emerged with a rodent in his mouth. What could have been going through Dark's mind at that time? Did he know, or even sense, what had happened between his mate and their son?

Big Guy disappeared from the region. I gave him up as a loss; after all, it was about time for him to disperse. I can't say for certain, but what may have actually happened was that Dark had told his son directly that he had to leave their home range and never return. The consequences of such actions—a son mating with his mother—are severe in our world, but did such drama play out in nature?

I was reminded some months later that foxes are not humans, that they live beyond our morality and thereby they are deemed wild. When I next found Big Guy, he was with the young female gray fox from Mama Bold's family, Pale, hanging out and paired up, over on Harbor Hill. They made a good couple.

One morning gray fox Brownie was determined to protect his territory from all trespassers. After all, he had a family over there at the ITT Facility in the middle of the Saltwater Wetlands marsh. I walked to the observation deck with him trotting alongside me. Down the road a fox, Blue, crossed through what I called "the junction," where another levee road intersected the main one: three choices. Brownie arched his back and tail, straightened and extended his legs, making him look far larger than normal. He bristled. He pranced about for a moment much like a horse or a ballerina showing off.

At that moment, Brownie wanted it known that he was the enforcer of his strip of the road where field mice filled his belly. He dashed at trespasser Blue who instantly turned for the brush.

I shouted, "No, Brownie! Stop!"

He froze, it seemed to me. He looked back as though to ask, "What in the hell do you mean, 'Stop' and 'No?'" By then Blue was nowhere to be found, and Brownie trotted back up the road to where I stood at the junction as dawn outlined the distant mountaintops with a needle-thin trace of light.

Foxes are capable of great hostility to one another. At the same time, they live on the edge of being social, at least to some degree. When the females in an urban setting hold overlapping territories in a pinch point, there comes that realization that there is no more room, and it is then, in such tight quarters, that natal den regions are forced to intermingle. The urban gray foxes have nowhere to go and so they get along with one another no matter what the season. In some ways, they become an extended family. Some are sisters or aunts, others are brothers or uncles, and there is a mix of ages strung out along the south side of the creek. Almost all of the foxes on the northside of the creek are unattached. Most of the time there is calm within the community. They tolerate one another. There are no turf wars as are found on the north side of the creek. There are no one-on-one fights, except when the alpha male, Dark, culls the gathering. I've seen him at work. I suspect that Dark Eyes, the alpha female in the region, directs everything he carries out. They are quasi-social, living side by side, rearing their litters, grooming each other, and nursing pups from the neighboring den, but never sharing.

It so happened that when the foxes had their pups out along the levee road on the northside of the creek, they lived beside the Black Creek Trail, a part of the Baylands public trail system. I'd been told repeatedly by the local ranger to keep my distance from the foxes, but there was a den of pups right next to a public area. I could not have asked for better luck, and I couldn't help but feel, in some small way, that the foxes had chosen the spot just so I could observe them. Their placement allowed me to feel like I had conformed to the ranger's dictates, but honestly, I stretched the rules by putting a trail camera down in the floodplain, exploring the north side of the creek, and going anywhere I chose, but I did so shrouded in what stealth a man of my age could muster.

During my observations, I picked up traits that the

foxes used to remain just out of sight. The foxes had already taught me how to read my environment, how to move without being known. I wasn't half as wily, but I felt like I used my fox lessons rather well. For instance, every time I emerged from the brush out into the open, I checked, then double-checked, to make certain no threats lurked nearby. A threat was simply a person, a stray dog, or just something that moved over there. I peeked. When I saw there was no threat, only then did I step into the open. I listened for voices. If I heard people talking, I made sure that they were over on the public side of the creek; otherwise, I froze, anticipating and watching as teenagers passed by in the channel. When a person or two passed over on the new trail at Harbor Hill, I remained still, or hung back in the brush under the canopy and listened as they talked and walked by. I hovered out of sight in the background. Looking back, I wonder how many people really did see me out there in the bushes, and what they thought I might be up to.

I knew that very few officials ever came out to check on me or see what was happening, yet I worked with extreme caution. I installed my trail cameras well off the beaten track, back in the brush, under the canopies, deep in natural clearings in the forest, and at the junction of several major animal corridors that, to the foxes, were the freeways of their lives. These roads and human paths gave them boulevards to trot along from one hunting region to the next, and as quickly as necessary.

Yet, there was no territory left for the gray fox pups to come into their own. There were increasing numbers of foxes on the move, looking for territory, but, in the end, when they needed to settle down, create a natal den, and give birth, they found that they had to live side by side with another pair of foxes doing the same thing. They had to learn to get along, learn to befriend one another instead of selfishly declaring their territory and battling over such a small area.

More pressing was the question, "Will there be enough food for all?" That consideration was truly contrary to the lifestyle of the traditional gray fox, such as Brownie, who, while his pups were still dependent, fought any trespassers who dared trot through his territory. Brownie and Little One, his mate, ruled the back road and the levee road all the way up to the gate by the Madera Creek parking lot. The fox families concluded that there was more than enough food in the region for all, but they needed Brownie's permission.

Brownie eventually gave in and allowed the others to come onto his home range. They were welcome and would be until his vixen was pregnant again, whereupon he would bite and claw any intruder until he drew blood from his opponent's thigh or shredded an ear. Foxes fight in a blur of screeching, squealing, and white-belly fur and tumbling tails, until one cries in submission. This often takes the form of the fox rolling onto its back and showing its vulnerable underbelly. Five to seven seconds is a long battle. They are stunning to watch.

This leap of faith on Brownie's part was the significant step into being an urban gray fox. A universe opened. It allowed the foxes access to dumpsters, to feed on feral kittens, to sleep on the warm tops of cars in the parking lots nearby. The gray foxes endeared themselves to the high-tech circles at the Los Palos Technology Center. People fell in love with them. I was asked to come in and speak about these foxes, to explain them to the people who worked there and were afraid of anything that even remotely resembled wildlife.

Wild conjures fear. Wild is uncontrollable. Some say that wild indicates an attack with the intent to kill. Does "wild" conjure such reactions in you, dear reader? Do you feel an aversion to the word, or perhaps some attraction? What does that word do within your mind and emotions? For many, it means instinctual fear. It reveals that, for far too long, we have clung to a misperception of control, that we are somehow not fragile, vulnerable creatures. We have armed ourselves with an arsenal of weapons to propagate the lie that we are anything but mortal. The walls rise higher, the fortifications intensify, and if something resides beyond the gates, it is *a priori* dangerous, for it doesn't conform to our frightened, sheltered, cowering way of life: our rules of conduct, our legal system, our civilization, our ideals, our social enjoyment of one another, our endlessly expanding territory. What we refuse to see is that nature controls us all, despite our ammunition, our poisons, and our braggadocio. By definition, "wild" is beyond our grasp, and we fear our weakness in its presence.

Is there another way to look at all of this? Science can learn from the past and posit the future, but at the same time there is room for science to grow; it is not the be-all, end-all. We are just beginning to recognize our lack of knowledge, just now changing our notions and beliefs about that which we call wildlife. Generations will pass before we learn the truth of ourselves and the world around us, before the slaughter of animals fades from human history. And it must fade in time, because I and others like me have opened the doors of understanding that cannot be shut.

THIRTEEN

TERRITORY BE DAMNED

Around April in the temperate zone is when fox pups are born. That's when territories become critical, when they're firmly carved into the landscape to prevent intrusion. The coming generation is that important to those foxes. They will fight unrestrained, causing severe injury to any trespassers, but there's never been a report of a gray fox killing another gray fox in a fight. It might still happen, but not from what I've seen.

Later in the year, in late fall, early winter, there are no territories, no boundaries that need staunch defenses, because there are no gray fox pups to protect and feed. Dispersal is underway. Each year, the foxes allow for a window when territories no longer need defending, when the borders drop and they are able to socialize once more. By dissolving boundaries after the first season, the pups can freely find mates and position themselves to raise a litter in their own territories, or in the urban setting next door.

In most cases, those first yearlings hunted for the opposite sex as diligently as they hunted prey. They sniffed the trails and the latrines to discover information about the surrounding landscape on their quest for mates. These latrines—places with multiple feces piled one upon the other—act as sort of library of the recent past, as well as a direct means of communication, for there the foxes can find such information as:

1. How many gray foxes of each sex have passed this way?
2. What food is available here, and how much of it?
3. **How many other gray foxes live nearby and how much territory is taken?**

These were some of the issues that faced the dispersing pups heading out into the world for the first time to live as adults.

Not all of them willingly dispersed, so in some cases

they needed a little encouragement. Some had to be discharged, forcibly ejected from the territory. Fearless, the alpha female of her litter, had not dispersed. She clung to the territory as if she owned it, as if it were hers by right. When her father, Ears, returned and found her there, he rushed her, growling, and smashed headlong into her. The force of his blow shook her. At the same instant, she understood what he demanded; she vanished.

I didn't see her again until the following year, when she came back through and had her own pups.

After dispersal, with such open terrain, with no territories, the foxes wallow in spirited polyandrous encounters. It is a time for play, for testing, and for getting to know each other. It is the time when Dionysus walks the earth and both female and male gray foxes strut. Every strand of DNA beckons them into the realm of the opposite sex.

This is when the gray foxes seek to bond. Females look for a male based primarily on how good of a provider he will be, just how deft he is when targeting prey. When his jaws slice through the flesh of a duck's neck, does he kill with ease? Does that suggest that maybe he understands pain and wants to minimize it? Is such an act viewed as utilitarian only, or is there perhaps an element of actual attraction, an appreciation for the beauty of death?

As for the males, they have to know that the female completely understands her role as a gray fox without a second thought. Does she perhaps have a light touch to grooming and favors ear cleaning? Does she love the taste of the bugs in those ears, especially the fat, juicy ticks? Will she make sure that her pups know the requisite memes—even the ones about human beings—such as what the family does when a raccoon crosses their path? Can she share their stories from a language based on body movement, in what we often call nonverbal communication?

For a gray fox, choosing a mate completely contradicts what most humans do; pairing up has nothing to do with sexual attraction. If the male catches the female when she is in estrus and she lets him in, the litter inevitably arrives. Out of a litter of five, one might be genetically his, but then again maybe none are. Due to the rapidfire transition of sexual partners during estrus, there's no guarantee that the "father" is actually the biological father. Regardless of whether he has a genetic relationship with his mate's pups or not, he still helps to raise them, still forms a patriarchal bond with them. He will protect the den and ward off trespassers.

He will come from off the hill with a ground squirrel sandwiched between his jaws, and he will deliver it to their litter to feed her pups, because they are his pups, genetic connection or not.

FOURTEEN

INBREEDING

Although they settled down nearly side by side, they yearned for the complete solitude of a vast home range with the guarantee of enough food for all. Living so closely together as they were then, along the creek, meant that there was no clearly defined territory, no clearly defined food sources.

Down along the creek, there is a huge pinch point, an area where everything in the wild comes together. It's like the tip of an arrowhead, a funnel, where the raccoons must learn how to live with the foxes, and the foxes must learn when to step aside, even when there is no room. Opossums came down the trail, stopping to wash their faces, stopping to tend to their young in the pouch, and the foxes could do nothing but watch it happen. No one left the den. No one left to go find themselves a mate or new territory because there was nowhere to go. The young foxes dispersed, but came back because there were no alternatives. They moved back in with Mom and Dad, or at least that's how I viewed the situation. Some dispersed, remaining away for a month or so, but they came back to their parents' home range.

Pups born in April of a given year mature about November, sometimes before. They were ready to pair up and have families of their own, but because they were born at the pinch point, they very soon understood that they had no choice but to breed with their brothers or fathers or mothers or sisters. Gray foxes aren't far-ranging, so they had to make do with what was available—in this case, family.

Incest was the norm along Madera Creek, and over by the water purification plant, and elsewhere along the San Francisco Bay marshes. There was no fresh DNA out there in the woods because there was no room, and there was no room because there was not enough territory to claim. The results were deformities like floppy ears, a genetic trait commonly understood to be caused by inbreeding. Were they due to an injury incurred during fighting, the edges of the ears would be ragged and torn, which I have seen. In her younger days, Tense was a fighter, and I saw her come away from a battle with a bloody, sliced ear from skull to tip. That's how I identified her.

Many opposed my viewpoint. Floppy ears occur in many of the gray foxes in the urban setting of the Los Palos Baylands. With ears like that, part of a fox's hearing is diminished, and since hearing is the single most important sense for hunting food to fill its belly and feed its young, such a handicap can have far-reaching, damaging effects on gray foxes. The real tragedy is that, if the floppy ear is genetic, each floppy-eared fox will pass this gene into the gene pool, weakening them until their immune systems are unable to cope with bacteria and viruses, until sickness overwhelms them, until the group fades away—skinny, fur bunched into ragged clumps of mange, injuries like infected eyes that do not heal. And soon, nothing will remain as canine distemper sweeps through the landscape like Ebola. In that stew of inbreeding, one by one, they died—and no one ever asked why.

I knew the reason because I'd seen it. The distemper raged because animal corridors were forever demolished, leaving the local wildlife in a prison of man's unthinking, ego-driven design.

FIFTEEN

RESTORATION

How will we reconnect the corridors around the San Francisco Bay? Given the human development of the area and the pinched animal populations, is there even enough room to rebuild habitats? How will rising sea levels affect the Baylands? How long do any of the creatures there really have left to live?

Before I could even think about how to approach solving those issues, the one nearest at hand was that Squat, the alpha male in Fox Hollow, had developed a pronounced limp. I couldn't discern the cause, but the impact on his life was apparent. The dilemma was excruciating: wait out the injury in the hopes that Squat would recover, all the while running the risk that he would suffer and very possibly die, or call for help from the animal rescue folks. I didn't want to interfere with nature's course, but I found I couldn't sit back and leave a fox in pain. I made the call.

"If he doesn't heal," said the good-natured young woman from their facility, "there are only two options for Squat: euthanization or placement in a zoo for animals that cannot return to the wild. Zoo openings are rare, however, so the chances of putting him there are very low. We'll probably have to put him down."

How absurd. How narrow-minded. I had reached out for assistance, and was instead met with an ultimatum. Imagine calling an ambulance for a sibling and being told they must be euthanized in lieu of actual care. The concept was outrageous to me. I would never permit Squat to be "put down," as they so euphemistically called intentional killing, and there was no possibility in my mind that he should be imprisoned in a cage for the remainder of his days. He was too wild for such inept handling.

I rejected all so-called assistance and let Squat limp for two painful months. Surely nature would run its course; you might say I had faith. Squat made do, and the rest of the family supported him in their own way, but we all need a little help at some point. I found some

119

in a colleague.

Greg Kerekes is the cofounder of the Urban Wildlife Research Project. Someone had referred Greg to me early on in my work with the gray foxes. Prior, there had been several people who wanted to be my assistant, but none had felt authentic. I had felt as if all they really wanted to do was see a fox, not study them, not be among them. When Greg entered the scene, I took him out along the trail, and he stood out as an authentic wildlife advocate; he had a wealth of experience that helped me to understand our subjects in greater depth. I asked him to come on board and he accepted.

One morning, Greg and I were up on Fox Hollow Hill looking down the back road. Greg had his big lens out and said, "There's a fox coming."

I asked, "Can you tell which one it is?"

"No, not yet, but it's not Squat. This fox isn't limping."

But as the unhindered fox drew near, I could see that it was indeed Squat, and he was no longer limping. The leg was still tender, but at least now he could trot. He was healed. Nature hadn't let us down. He strode past us with an elegance that no other fox could have conjured, and told me in no uncertain terms that he was whole once again.

In another health scare, Midget turned up with a

slashed eye from one of his many fights. It looked ugly; pus drained down the side of his left cheek. He was only three months old at the time. He healed within a month. With that, I came to trust that nature would heal all wounds, until I came up against Gray's eyes.

Gray had developed an infection in both eyes. I first noticed the ooze from his right eye, which grew into a crust that blinded him. However, he still had his ears and nose to guide him. One of his eyes cleared up. With one eye clear, he dutifully went into the marsh each night and delivered three or more rodents to his five hungry pups. Gray was a master hunter, even with his impairment. Few foxes could so consistently succeed in tasting their prey as he did.

I watched it all unfold over the course of a year. The foxes ate certain grasses perhaps as part of their healing process, but I couldn't know for sure. I'd see grass in their scat mixed with worm eggs. Invariably, in the region that I monitored, the foxes suffered from itching, some more than others. I trusted nature to heal until I saw that there was no progress, no cessation of that ooze from his eye, no healing, no relief for the master hunter, Gray. He alone suffered.

I decided that I'd make this an exception, so I sent out a query for people in the network to come and take care of Gray's eyes, to heal him. We tossed together a team: a few vets who knew wildlife, a permitted trapper from the California Department of Fish and Wildlife, and me, the Fox Guy. All the bases covered. I helped the trapper, Terris, position her traps, knowing that Mama Bold and Gray were then hanging out back by the big acacia bushes between the truck washing pad and the water purification plant.

Terris said, "Call me when you know we got Gray."

By the next morning, we already had a catch. I called it in: "Young raccoon in live trap number one."

This went on for some time as various wildlife wandered into the traps, until we finally snagged a fox. When I got there and turned the trap so that I could see which one it was, there was Mama Bold, still and calm as always. She gave me a look that I could only interpret as, *You helped them catch me, fox human?* I had indeed done just that, but I reminded myself that this was for the good of her mate. Terris opened the trap and Mama Bold shot through the tall grass dead ahead. Overwhelming freedom consumed her as she fled, to the degree that I feared she might run headlong into a tree trunk, but she dodged them all and disappeared.

Twice, we caught Mama Bold in a trap. The second time she fought to be free, and the cage rattled from her urgency. When Terris opened the trap, Mama Bold exploded into the underbrush with reckless abandon.

Gray avoided the traps, but I saw him shambling about the area as his disease progressed. I watched his eyes grow covered with a thick liquid oozing from the corner of his eye—an ocular discharge, as they call it—and he used his dew claw to clean the dry crust that blinded him, but his attempts were insufficient.

As days passed, and the traps failed and were withdrawn, I feared for him. There was still the possibility that one of the vets might notify an animal hospital. Would the professionals call for his euthanization, or his imprisonment? I thought of Gray—the alpha male, the master hunter, the father—locked into a space no bigger than his body, removed from the familiar trails of his home, removed from his family, left with only his memories of Mama Bold, and I shuddered.

He'd suffer confusion at the very least, for he'd lose access to the familiar sounds and smells that had mapped out his entire world since birth. None of the usual tree branches would be there to guide him through the brush on his way out to the living marsh, nor back again to the family den. All would be lost, incomprehensible to his mind. Drenched in fright, he would be unable to even form a question about what had happened to him.

I had to do more, do better, to spare him from that fate. Gray needed medicine that only humans could provide...didn't he? I felt that he had somehow injured his eyes, that possibly there were foxtails or some other organic matter buried in his eyelids, causing the infection.

I doubled my efforts to find him, called for Terris to reposition the traps again and again, and stayed out far later than usual to catch a glimpse of the miserable alpha, but to no avail. Gray was nowhere to be found. I feared his blindness would be permanent, his quality of life forever lost.

Then one evening, Mama Bold came to me. She approached from the big eucalyptus near the end of Fox Hollow Trail and came right up to where I stood.

Follow, she told me, and so I followed.

She led me to the alkali saltbush, and there, at the top of Skunk Hill, was Gray.

And his eyes were clear.

But how? How had a disease so debilitating suddenly disappeared? As I walked toward Gray in awe, a third fox crested the hill; it was Pale, their daughter. Pale drew near to Gray and made to greet him, but this time, she didn't go through the ritual of submission by coming up under his chin. Instead, she lay down before him, exposing her underbelly. In return, Gray nuzzled Pale's jaw, licked her, then fell into a full neck rub, scent marking her with his tongue as he went. He rubbed her neck with a liquid stroke.

Pale understood the gesture, but I couldn't believe what I was witnessing. The neck rub signified total pleasure, a deep gratitude, and an expression of recognized value. I'd seen them neck rub objects they'd enjoyed, such as the grass or my handkerchief, but never before had I seen any fox grant another such a gift, especially not an elder to a juvenile. In the case of father and daughter, it was miraculous.

Gray's capabilities were amazing. He was a master hunter, a defender of their territory, and a fighter like none other, and at the same time his heart melted when he played with his pups. He gave them their identities and taught them well—taught them what it took to be an able gray fox.

Why would Gray make such a magnificent gesture to his daughter, unless she had done something extraordinary? Was I projecting a story where none existed, an arrogant human interpreting this scene as anthropomorphic? I didn't want that to be the case, but it was certainly possible. I've caught myself unconsciously projecting human qualities on creatures more than a few times, I admit, but I strive to see these foxes as they truly are, not as I wish them to be.

Gradually, I realized what Mama Bold had wanted me to see. My eyes were opened as surely as Gray's had been. We saw, Gray and I, the tall dry grass beside the trail to Skunk Hill. We saw his weeds, his alkali saltbush off to the right, and over where the big eucalyptus trees stood, just as they had been before our blindness.

I concluded that Pale was responsible for her father's clear eyes. She had sat beside him, licking his eyes, licking away the crust that had darkened his sight. She'd licked until his eyelids fluttered. Pale had licked his eyes and washed them clean with her saliva, that naturally curative substance from her own body. She'd given him sight once more, and that neck rub said it all: how thankful Gray was to have his sight back and how thankful he was that his daughter had made it happen.

Gray walked down the hill, over toward the eucalyptus trees. Pale dashed ahead and sat on one of the branches of a fallen tree. Mama Bold followed her mate. When they came to the tree that Mama Bold had nearly run into not one day before, the pair easily slipped around it to sit on the other side. Gray was restored, once again so aware, so in tune with his environment that he melted in between the cracks. His movements flowed like the river.

SIXTEEN

DARING: THE FALLEN ALPHA

I remember that little female, Daring. She came from one of the litters of five that, accompanied by Gray, simply came out from the grass and brush and sat there on the road looking up at me, wondering what the hell this thing was that stood there looking back; her mother, Mama Bold, had done the same when she was small, as curious about me as I'd been about her. She hadn't yet lost her pudgy pup nose, but she had taken on some of the adult color. Daring's four siblings watched from behind the thick brush along the chain-link fence near Fox Hollow. They didn't understand her, but she was their sister and that must have meant something to them. Gray foxes seem to establish a bond within their family.

When someone came jogging down the road by the water purification plant, all the other pups dashed breathlessly back into the brush. Not Daring. She stood there tipping her head from side to side, trying to build an understanding of whatever human being crossed her path. I sensed that she embodied serious power. I knew that she was the only pup in the den that would naturally emerge from the brush, overcoming any fear, because, after all, the adults had no fear. And why should they? I was harmless, and the runners were either too winded or too fixated on their running postures to cause them concern. That was all they knew of humanity. Since her exposure was so limited, she was fearless.

From the beginning, Gray saw exactly which one of the pups he would nudge toward being the alpha of the litter. Daring grew into her role very quickly. Her status was clear:

> 1. She had first choice when any food arrived, and none of the other pups had the right to eat until she was finished. In a den of five pups, there was always one that ate last, and sometimes that last pup died due to lack of nutrition. The alpha's needs came first.

2.She was the one who had the right to dictate the lives of her siblings—their comings and goings, or even the places they sat or played.

3.She learned to use violence to maintain order among her siblings.

From the time she could begin to make sense of the world, Daring would lay on her back with her ears flat on her head. She bared her teeth and growled at her four siblings, keeping them in place while at the same time submissively favoring her mother, Mama Bold. Her father, Gray, came second in Daring's eyes. She showed her teeth whenever it suited her, and her siblings all obeyed.

Daring didn't have to try. She didn't have to work to attain her status. It all flowed naturally, like the sacred and mighty Flathead River threading its way through the landscape for millennia. Daring came into this world with "destiny" written all over her. Whenever her siblings came to meet her, they slipped their bellies to the earth, cocked their ears back, and made happy, squeaking sounds in a performance of submission for their preeminent sister.

Daring had but a single flaw, a blind spot: she thought she had to fight to maintain her status. She couldn't believe that just being herself was adequate. Insecurity caked itself like hard clay around her ego. She could not see that something so natural needed no defense. At every opportunity, her aggression grew, and she began snapping at any fox nearby. She pulled her head back, cocked like a hammer, bared her teeth, and growled. She ruled through tyranny and wrapped herself in a blanket of brutality.

I once watched Gray carry a Canada goose back to where the pups napped. He managed to dump it in the clearing, exhausted, before Daring started her feast. Later on toward evening, he dropped the still body of a woodrat in the midst of the five pups, then walked off to the side and was asleep within moments.

Daring dashed for it. She positioned herself between her siblings and her prize, clarifying that this was hers alone. Her lips were pulled back, exposing her teeth; the rest needed to wait their turn. Once Daring had had her fill, another took the carcass, and another, and so on until it came to Shy's turn, the last of them. There was very little left on the bones of that woodrat by the time Shy got to eat. Nonetheless, she was grateful for something.

Shy most often went about her day hungry. She chose to lie back on the hill near the alkali bush natal den. The rest of her siblings chased and wrestled, energetic and vibrant, down by the abandoned den along Landings Road.

Daring held the power, and she was out of control.

On one particular night, the feeding time did not go as expected, did not take its natural flow. Shy lay on the dusty brown grass. Now and then, hunger cinched her stomach. Before Shy realized what she was doing, the squirrel's leg stretched between her teeth and Daring's ravenous backwards pull. Instantly, alpha lunged at omega. There was no time to show remorse or take back her actions, so for the first time in her life, Shy's gnawing hunger caused her to react with force, squealing in fury as she buried her teeth in Daring's hind leg. A split second later, she chewed and bit Daring's ear until it went limp. Pain surged like electricity throughout their bodies.

Daring's hind leg gave out, falling limp as sharp charges of agony bolted through her. Daring froze in mid-stride. Again, Shy bit Daring's ear, shredding the cartilage. The ear no longer stood upright, but folded over and impaired her hearing. Ragged, beaten, and bewildered, Daring left the prized meal, as well as the den, and slunk off into the tall grass nearby. Her leg burned. She licked the wound and tasted blood. First blood. Woozy, lightheaded, she lay on the rough grass, allowing the pain to envelop her in defeat.

Injury to the hind leg of a gray fox is a serious, multilayered handicap. It carries the failure of prestige no matter where on the hierarchy the fox might fall. Daring's plummet from the top began with Shy's attack, but it continued with her sibling, Slim. No longer did he greet her with a fox-kiss, nor would he slither up to her to nuzzle beneath her chin in a submissive gesture. Whenever they met, he shunned her. It took but a single night for the others in the litter to follow Slim's lead.

Daring, the once-alpha, was systematically demoted, and the respect that her siblings once gave was whisked away. They no longer approached her, nor did they show any deference. They simply ignored her because she was, in their eyes, broken. Her damaged leg prevented her from climbing a tree to escape a coyote, or going far out on a limb if a bobcat followed. Her mangled ear restricted her hearing, which was an impediment she could not overcome. She was a burden, embodying weakness. Being handicapped in such a way meant that she was no longer the alpha female pup that everyone looked up to.

With Mama Bold watching, Daring tried to rise and take a step, but when that hind leg hit the dry weeds, a sharp pain shot through the pup's body. Her leg kicked and buckled. Fire gripped every fiber. She stood trembling in agony, but made no sound. She had to move, but with an injury like that, she could only stumble. On top of that, her right ear flopped over onto its side, making it impossible to use for bidirectional hearing.

Daring could no longer climb trees, because doing so required both hind legs to shove her up the tree's trunk, and she only had one good leg. She hugged the trunk, desperately wanting to climb like her brothers and sisters, to lead them far up into the great eucalyptus trees, but each step shot lightning through that thigh. She could hardly walk, let alone climb.

Mama Bold, watching from nearby, saw Daring trying to lift herself up that blue gum tree to join her siblings in racing around through the branches in a happy frenzy. The pup slid from the tree and stood there in defeat as Mama walked over and briefly nuzzled her. It was just enough to let Daring know that Mama still cared for her broken daughter.

There is a brief period of maybe two weeks when the pups bring to the den whatever they've successfully killed. It's essentially a time for the would-be hunters to prove themselves. Daring's injury, however, meant that she could not hunt, could not bring a ground squirrel or woodrat back to the den. She could no longer support the den and became a dead weight and a drain on resources. She became a leech within the gray fox community. Before the injury, she could crouch, position herself exactly, listen to every movement, set her taut muscles to launch her body into an elegant arc, and come down on a field mouse with a precise snap of her jaws. No longer could she do so.

She was less than a gray fox, laboring under severe depression, moving slowly and carrying herself slumped as if beneath a heavy burden—an impossible load. She suffered, and yet she tried everything she could think of to regain her stature. She played at being kind, but everyone saw through the ruse; they knew she wanted to be back on top, and they knew the dangers that would entail.

That hind leg kept her hobbling along. It stopped her from climbing and from hunting out on the marsh. Daring was an outcast for far longer than she could tolerate. Weeks ran sluggishly by. She took the leftovers of the duck, the rat, and the squirrel. She lost weight. Sometimes she went days without food because Gray dropped it off for the others to eat before she'd even come to the den. There were times she couldn't even crawl that far. Her stomach quavered and tensed.

Her life seeped into existing as an outcast. She eventually stopped trying, sleeping on a soft pad of grass most of the day and into the night out on the edge of the vast marsh.

One afternoon as she awoke, as is usual for a gray fox in midday, she stretched her hind legs—something she had been unable to do for quite some time. That injured leg extended, which hurt, but it felt so good to move those muscles. She stretched, putting more and more tension on it, until a jolt shot through the nerves of her thigh. A surge of confidence twitched in her. From that moment on, she knew that she had tapped into the healing flame.

She licked the wound even more than before, keeping it wet with her saliva. She stretched until pain knifed through her, then backed off, rested, and tried again. Day after day, Daring could do a little more and a little more, until she could walk without limping, then trot down the road at night without stumbling. Then she listened in the grass and realized she could hear the field mice run. Finally, her hind leg felt strong enough to put to the true test.

Alone, she stood before the old climbing tree. Without forethought, she hit the tree moving, her hind legs pushing, the pain in her thigh biting, but she kept pushing and grasping and pushing until she rested on a branch high above, breathless and elated. Every night for more than a week, she climbed that tree. Each time there was less and less pain until finally, as she shot up the tree, the pain abandoned her.

It didn't take long for her sister, Shy, to come around and sniff. Back in the clearing, she came from the blackberry thicket and sat not far from Daring. The others watched and knew that their sibling had regained her strength. Within a day, her family no longer shunned her. They saw her as an equal once again.

One night, as a crescent moon hung overhead, Shy approached her sister and lay flat out on her stomach, ears pasted back on her head, her bushy tail with that signature black tip swishing in spacious arcs across the ground, kicking up dust, as she twittered and squeaked and yipped—a gesture that goes back millions of years, long before a wolf knew what it meant to be part of a pack. Swishing is the ancient antecedent to tail wagging, and it's still displayed by the gray fox every day between subordinates and alpha foxes, or between friends, siblings, or any other compatible foxes.

That gesture has meant the same for millions of years. Every time I see it, I know that I am seeing into a past that is far older than humankind, deeper than I can even imagine. Everything about Shy's movements exuded sheer happiness, sheer pleasure. There beneath that moon, Shy performed the ritual that brought Daring back into the fold as the alpha female of their litter.

SEVENTEEN

MY MISTAKE

I turned my attention and my cameras back onto the play out in a field of sienna-hued grass. The pups chased each other across the grass; they wrestled and played a game of attack and defense, which would teach them how to survive. Still, there would be casualties.

Midget growled at Tense. Shrieks and pitched gasps burst in a flash, an undertone of growl thrummed, and the fight stopped. Tense left, happily swishing her tail. She went to her biological alpha female, Dark Eyes, where she sat watching on the slope. They met and exchanged a fox-kiss, then Tense rushed back into the dry weeds and vanished into the thicket.

Somehow they knew the coming of day—the pattern

lay out in burning red in the east, and the sky, filled with fire and color, changed so quickly as dawn took hold beyond time, giving birth to the sun shining everywhere. What an exhilaration to experience just how fast this ball of mud called Earth spins, as this point in time gathers my life along the levee road, as dawn scours darkness from the mountains. Earth has created another day.

Sure enough, there Dark Eyes came, trotting directly at me. She stopped just in front of the gate and lay flat. It was light enough that I could see. She turned and looked behind at the road, back in the direction from whence she'd come. Instantly, I saw him, a black pocket of power racing through the dim light. He shot from off the landfill with something even darker clenched between his teeth, dashed beneath the gate, and flew into the brush on the other side, where the natal den lay buried in the tangled thicket.

Silence hung in the foggy air for a moment, but then the pups yipped over food.

She stayed at the gate, lying on the dirt road, intently staring at me seated there on my stool. I jotted down notes as she barked in that hoarse, sandpaper rasp they deliver when they are seriously upset.

I tried to imitate her sounds.

She "barked" back.

We exchanged barks back and forth.

I knew that she shouted, "Get the hell out of my home, or I'll do something rash."

I barked and she came straight at me, rigid with a certain fury in her eyes. Fur stood on her back. Fright shot through my stomach and into my chest; I gasped. That was the first and only moment in my life with the foxes that I wanted a stick—something to lash out with—because I was certain that that fox might very well attack me. In that instant, I wondered whether my jeans would offer any protection from her teeth; I didn't like my odds. I imagined those teeth knifing into my thigh and felt imaginary pain at the very thought. My eyes darted from the fox to the ground and back again. I really needed a stick.

She stopped and barked, furious at my presence, then dashed for her den and her pups across the road. I let loose the breath I'd been holding and heaved a sigh of relief. Sweat beaded on my forehead and the back of my neck. I realized I'd been clenching my stomach in primal fear, and it took some minutes to realign some illusion of safety. I'd walked a thin line just then.

Only then did I realize that her pups were more important than dealing with some human trespasser on a stool. All she needed was to destroy the barrier between her pups and me. I slowly inhaled, thinking about what had just happened. I turned and walked in a circle, taking stock of what I no longer knew.

That was one of the last days I saw any fox in Fox Hollow for the rest of that year. When they vanished, I hadn't yet learned enough about them to know how to move with them, much less whether they would return. Questions bloomed and I blamed myself for chasing them away. I thought their dispersal was my fault.

I had yet to know even the rudiments of their behavior. I had no idea that there were other foxes around. I had stumbled, and it would take time to learn. I wouldn't understand dispersal for another year or more. They were there one morning as if nothing was out of the ordinary, and the next day they were gone, and the next, until they'd been gone for weeks. I had no idea where they had gone, and I didn't know if I'd ever see a fox again. They had left me all alone in Fox Hollow, and my dilemma surfaced: I wanted to understand these gray foxes, but there were no foxes to study. Fear gripped me, and I worried the loss was permanent.

I'd made a mistake.

EIGHTEEN

A PRELUDE TO PERCEPTION

Compared with the gray fox, we are deaf and can only smell a minute portion of the world happening around us. We evolved through a different path. We can only bridge that chasm through endless observation, I suspect.

The foxes slice and shove through a fog of odors to find the truths of their environment. They perceive through their highly refined physiology of smell, and some canines have up to three hundred million olfactory cells in their sniffer. The scents flow into two separate channels in the first cycle, but then those smelly airwaves, those odiferous atoms already sensed, enter a second set of olfactory nerves even *before* the information reaches the fox's brain and is acted upon. It all happens in a fraction of an instant.

You and I have sixty thousand such olfactory cells and a single channel that can get all mucked up with snot and other impairments. We cannot compete with a fox's power, much less understand it. They are in another universe, another paradigm.

They rely 90 percent on their hearing when it comes to food, and they each catch a rodent or two a night. The scent in the air—crucial to understanding what is going on and who the players are—makes up another 5 percent. My estimates come from observation alone, but they seem correct to me, and for me that is sufficient, at least for now. The other senses play a role in rounding out the percentages, but they are minor by comparison. The sum total is the way by which they mold their universe, and by which each fox has its own technique to carve it out, its own foxsinality, its own way of being. Each fox expresses itself independently. I have seen it.

Gray foxes' ears are nearly constantly moving, scanning their environment like two radar dishes sweeping the landscape. For instance, they might point one ear behind and the other off to the left, calculating how distant the source of that sound is, its direction, whether it's moving toward, away, or across, while at the same instant deciphering any unrelated sounds, such as assessing a threat from over to the left or an innocuous movement behind.

If you watch the foxes' ears, they read their environment in nearly 360 degrees without turning their heads. Their eyes do little at this point. Multiple sounds translate in their brain, and at the same time they automatically make sense of it all. Neurons like ours talk. They read their environment through sound even more clearly than we read our environment through sight. The only exception to this is when they need to be keyed in on a single event, like a field mouse running under the dry grass.

Then both ears are rigid, focused, pointing in the same direction. They process bidirectional hearing without even a thought. It happens in their brains like a flash of phosphorus—two critical pieces of information understood instantly. That's just the way they think, and I've seen it repeated countless times.

As I watched the foxes, I saw that they moved their ears. I watched as they entered a new region and sniffed the air, their noses poised upward at a near 45-degree angle, picking up on which woodrat, which opossum, which raccoon had since been that way and how long ago it had passed through. They sniffed and heard their environment in real time.

The fox uses its ears to think, and, secondarily, it uses its nose for detail. How can we hope to understand these gray foxes when we don't even function in the world through duplicate sensory organs? We think we do; we all see, all smell, all have the same single set of sensory organs, but when applied from species to species, those organs function so differently that there's almost no comparison.

For us humans, smell is the most memorable, but least recalled, sense. Though our epistemologies are radically different, humans and foxes might someday come to understand each other. By that I mean, let's try to step into each other's spaces, trade perceptions, get a glimpse of the other's worldview. Let us as humans see through their rumbling growls, taut muscles, thrumming hearts, and more.

When we try to understand them through their tools of perception, we grasp the environment from a radically different viewpoint. We learn to listen, to hear, to smell beyond what we have known to this point. What is it like to be a foot and a half tall, able to tighten down into a slim parcel and move through the underbrush, moving through places that we humans could never dream of following?

Walk around one of these days with your eyes covered. Can you get through your day on smell alone with that old nose trying its best? Moreover, what about your ears? We human beings living in a city culture, a town culture, a regional culture, out there on the land away from any easily accessible McDonald's counter, cannot get around without ample assistance.

The other day, I came upon a sleeping gray fox pup. It kicked its hindquarters and twitched, and its left ear scanned. It lay on a bed of dry grass. The pup dreamed. Those who own dogs have seen this happen. There are a couple questions here: was that dream filled with visual imagery like mine tend to be, or was that dream constructed from sound and smell alone, with a shadowy sibling in the background?

As a person, I find it difficult to climb into those senses and understand their world. Smell is food at a well-appointed restaurant. Sound is a discussion. We have simply evolved using different modes, different tools of perception. We are no longer wild. As such, it may be futile for me to try to understand them, those gray foxes, because they are of another paradigm that has not yet been plumbed.

The colors they see are far different from the colors that flood my gaze, for I was born seeing three primary colors in all possible combinations, while foxes were born with only two primary colors. I read those numbers and am tempted, but I shouldn't be too hasty to label such as a deficiency. Do they see me more vividly when I wear my yellow shirt? As I change clothing, my colors change. Foxes might think, "It changes fur at will?" What might they make of that? Might they believe that evolution dictated my capability? We have evolved as beings who rely on our eyes, on colors, on shades, on angles, but we are far from superior in that arena. There are colors around us that we do not see unaided.

Foxes don't need to use their eyes as we do, and therefore, evolution has simplified their visual cortex but expanded ours, though only by one degree. We see in daylight. That's the way we've evolved. They see best at night through low-light infrared filters. Colors affect them differently, and rightly so, because their eyes have evolved to see in the night as clearly as you and I see in broad daylight, yet the images are different in substance and scale. They have adapted to low-light conditions, yet can still see in both brilliant day and darkest night.

How much color do I need to give shape to my world? For them, the shirts I wear will never be red. Red is impossible for them to grasp. If we sat down with a gray fox and talked to it about the color red, it would express nothing but confusion—an outright failure to even grasp the fringes of understanding red. To the fox, the concept of red would sound like hallucinogenic magic. It is beyond their knowledge.

The gray fox does not so critically need its eyes, as we do, to navigate, to feed, to raise its young, or to grasp its environment. Therefore, it does not need a varicolored view of the world. It evolved based on sound and smell, and that is more than sufficient. All else is supplemental.

NINETEEN

SENSORY TOOLS AND LONGEVITY

We can only engage our world using what sensory tools we possess, the very ones we use to construct our universe. For the gray fox, hearing and scenting snuggle very closely together and work in tandem. Given that neither fully dominates, since hearing and smelling are both technically primary, might it be that there is a language buried within an odor or a rasping bark, and within them, some specific meaning? How can they know things by smell alone, like when Dark raises his nose a bit and sniffs off in my direction? He knows me by smell, and by sound when I talk to him alone, or when I grunt or belch or fart. I have a signature and he knows me, at least in part. I know him by certain traits, too.

Dark's left ear was mangled into a circle, most likely in a furious battle with another fox. I identify him by the shape of his deformity, as I do with so many other gray foxes; they are no strangers to violence.

All of them, once you come to see them clearly, have unique signatures. Finding them takes time. We know that scat contains multiple messages. We know virtually nothing about the meaning of odor to a gray fox. The odors flowing on the breeze must be infused with information we cannot grasp.

Is there also a language buried within their ears, in their understanding of sound? Is information transmitted through what they hear? How is it that sound means the same from one fox to another? Are they capable of assessing the meaning of sound, or do they "simply" react?

When they are hunting and hearing their prey all about, how does the gray fox determine which woodrat will fill its belly? How does it know how far it must bound into the weeds when the field mouse scurries about beneath? The mathematical solution is found only in esoteric math. Is it calculating the amount of pressure, the degree of spring it must exert, the exact tension it must release from its hindquarters, the exact angles needed in order to kill, or are such calculations innate?

In any case, the questions we ask must remain clear, unmuddied by projection. We must not confuse ourselves and inject our humanity into the picture.

How is it that these gray foxes, these *Urocyon cinereoargenteus*, are the most basal of all canids? They are older than the wolves, the coyotes, the jackals, and all other canines. They had millions of years to evolve and learn. Why are they not more advanced—or are their advancements just more subtle than we can perceive?

What patterns of being are required to survive for all those ten million years, as some argue, for the gray fox? Why is there such longevity in some species and not within others? We've already taken a look at their sensory systems, like smell. The following section is supposition, but it seems to speak of their survival, of those elements ingrained in longevity.

TWENTY

SO, WHAT ARE FOXES?

They are only a bit over eighteen inches from the ground to the top of the ear. These gray foxes live in thickets, in places where blackberry vines wind and twine, choking all, where willows hang over the creek's bank, where the tall cottonwood measures the sky and bushes, and trees like those giant eucalyptus trees, dead and fallen, lie in underbrush twisted and tangled. They do not live out in the open like the red fox. How do they know how to quickly negotiate all of those possible ways to get from A to B? They listen to the multitude of sounds emanating from across the landscape and mix them with subtle odors. What does such a construct tell the fox?

Because their hearing may construct four-dimensional audio models of what's going on around them in that field, they learn to understand in a spatial capacity rather than what's merely apparent in front of them. I have no idea what the sum total of that information might be. It's difficult to even imagine that style of understanding, let alone what it would be like to possess it, to rely upon it. Then toss in the old nose, and what a potent mix of inundation the gray fox must experience at any given moment. What texture would that knowledge be? What flavor? Does it feel like music to their minds?

The auditory is primary (technically with olfactory, as I've mentioned previously), then the olfactory puts substance over sound, and in a distant third, not close, is touch. *I don't like when my paws get wet. The ground that I tread is simply covered with wet, so I must step lightly. Once in a while I come crashing down on a sharp pebble, and ouch!*

This is the way that they construct their reality, or so I theorize. These are the gray fox's tools of perception. Granted, there is a lot that I know nothing about. What they see and hear and smell beneath the brush, along the animal trails that are like freeways to them, limits my understanding, because I cannot imagine being so small, so enveloped by plants and trees, yet knowing my way back home without fail. How do they navigate?

How do they know where their destination is in that jungle of thicket? How different is their environment from what we experience, and what does it "look like" to them? I marvel at how powerfully we delude ourselves into presuming that ours is the only way.

If I think in philosophical terms, these gray foxes are existentialists. The moment is all that they have, but there is a latent part of every moment that allows for true intelligence; in their world, the moment governs, and yet they have a sense of time. They are able to anticipate, to plan for the future. They are able to assess just how much territory they will need to mark so as to feed all of their young. I have seen signs that they more than likely have an eidetic memory. They never forget.

Kingdom	Animalia
Phylum	Chordata
Class	Mammalia
Order	Carnivora
Family	Canidae
Genus	*Urocyon*
Species	*U. cinereoargenteus*
Binomial name	
Urocyon cinereoargenteus	
(Schreber, 1775)[2]	

In such classification charts the gray fox seemingly was unknowingly shoved off to "Order: Carnivora."

Why not assign the fox its own family? Since it is neither canine nor feline according to my own study of their behavior, the gray fox needs its own classification, given the current scientific rubric.

They have been misplaced.

TWENTY-ONE

FLEXIBILITY— PATTERNS IN MOTION

For many creatures, their habitats are doomed by human expansion in an attempt to accommodate the increasing billions of people. They are shot from the sky, or outside their burrows, or beside the watering holes for food, for sport, for nothing. They are hunted for supposed cures to diseases, ailments, and erectile dysfunctions. Some species vanished almost as quickly as they'd arrived, for when change swept their landscape, they stumbled and lay sprawled in the dust, marking the end of their kind. They died from want or flood or famine, and no one cared, much less knew the reasons why.

Some animals have special diets for the kinds of food

that they find pleasing and nutritious. The gray fox eats almost anything and in varying stages, from fresh grass to cached meat that's just turning with a jellied, translucent coating that preserves the meat and keeps it edible. Unlike with our food, maggots do not invade the meat of the gray fox. What preservative exists in their saliva? I have not investigated this, but it seems as though someone with the right equipment needs to take a look. Most importantly, their taste buds are such that there is almost nothing that they reject. Lizards, snakes, beetles, grass, burrito juices on aluminum foil, ducks, geese, and all flavors of bloody rodent. What a gastronomic variety.

One evening, the alpha male, Gray, came up on Fox Hollow Hill. Not far from the den, he walked carefully, slowly across the hilltop, went to the base of a eucalyptus tree, dug for a moment, then pulled from the earth a jackrabbit's thigh, coated with a jellied layer. He pulled off strands of meat.

I looked at the jellied coating and wondered how it worked—how everything worked. What mechanisms operate within these creatures that allow them to perform miracles? The fox adapts to droughts by knowing which plants contain water. The rodents that it devours contain liquid, and that adds to their supply in the Sonoran Desert. Their evolutionary bodies are like wanderers. I have seen them fully heal from what looked like serious injuries. I have seen them playful beyond words, and depressed, shoved down into the dark pit. I've seen them kill, and forgive, and kiss.

Beneath all that, the gray fox is a keystone predator in the urban forest. It always lives in a riparian habitat and abhors open spaces but it will tolerate some as long as shelter is nearby and there's a downhill run to boot. Its presence is so often overlooked, yet its impact on the world is undeniable.

Under some circumstances, if the gray fox alone were eliminated, it would cause a trophic cascade. That's where food becomes scarce or simply unavailable due to overpopulation of subspecies. If too many herbivores overwhelm the grasses and other organisms fed by photosynthesis, the environment will diminish and possibly collapse. It will no longer support life in any form, simply because the foxes are no longer present.

The gray fox knows how to exist between those cracks in time and balance out the weeds, the trees, the opossums, and the skunks.

In some ways, flexibility and adaptability overlap, but one difference is that adaptability underlines the issues of life and death. It is the physical day-to-day labor by which they live. They must be flexible so as to persist. Does the gray fox's physical structure determine how it survives? Climbing trees can save it from predators, while at the same time, the fox that climbed the tree can eat the dove that perched there for the night.

When a gray fox needs to use its eyesight to assist in finding something, sight merely augments their senses of smell and sound. It seems as if they have such a long muzzle that it blocks a portion of their sight. They may have a large blind spot, just as we have a blind spot between our eyes. As such, they can use their eyesight to see things on the ground some distance off, but they cannot see up close. That blind spot is rather large by my reckoning. They do not have true binocular vision except at a distance.

Touch is simply a utility. As far as my observations go, the only time that they pay much attention to their paws is when they bite to remove burrs or stickers, and when they have to cross water. Like cats, they do not like to get their paws wet. As with all else, I have no proof, but it seems to me that touch is only important when the pups are very young.

These are some of the abilities and characteristics of the gray fox—characteristics that have allowed them to survive and evolve for ten million years, according to what we presently know.

TWENTY-TWO

A BRUSH WITH NATURE

Stanford University had recently moved a large portion of their managerial staff to the Los Palos Technology Center, and their offices were settled at the edge of a vast saltwater marsh. Wild animals lived there.

People unfamiliar with wildlife of any kind fear the timid gray fox, for they only know that it is a wild animal, and emotionally for them, "wild" is aggression, attack, blood, pain, danger, death, and self. For them, "wild" is shrouded in a fogged understanding. They stand and say, "Excuse me, the only thing I care about is not getting bitten and infected with rabies, thank you very much." Behind that fabrication arises an ancient sense that anything wild must be an adversary, some predator to be eliminated.

I have felt myself embedded within that hunting party, living apart from the wild and driven by unending fear, and I now renounce it wholeheartedly. Not everyone sees eye to eye with me, but hope stirs in unexpected places.

A woman from the university said, "When I got this job, I was warned that I might go out to get in my car and see a gray fox sleeping—you know, like on the hood of my car. I did, and I thought it was strange at the time, but now I see that it was just having a nap."

Stories swam through the offices. They had never encountered a wild animal in their life until a gray fox lay sprawled out on their car's roof. The office workers named him "Sleepy." Many assumed that Sleepy would be aggressive. If not now, then soon…sometime. But all they found was a lounging fox. He endeared himself to everyone in those high-tech and international law offices there in the Los Palos Technology Center, out on Landings Road.

Sleepy turned people inside out, meaning they were touched by something they couldn't identify when they saw a wild urban gray fox pass by their office window. The woman said she sat in her office looking out into the parking lot. A fox roamed, free and unperturbed, but also non-threatening. "Just there," she said. "Just outside the window. If I tried, or wanted to, maybe I could reach out and…"

That sleepy little bush dog gave the people a break from routine, gave them the rare opportunity to encounter live, vibrant nature, just for a fleeting while.

And not one of them died from rabies.

Suggesting or saying the word "wild" taps an ancient and sometimes dormant part of our psyche. I can certainly imagine myself living fifteen thousand years ago, and although we preyed on certain others, we never preyed on our own. I can see we humans innovating, adapting, and learning how to defend ourselves from predators. Some have proposed that the reason we learned to sleep was to avoid being detected by nocturnal predators, and that makes a frightening bit of sense.

But before the people got used to seeing Sleepy and the other foxes around, some of the workers were frightened. The facilities manager contacted me and asked if I could come in and talk with the staff about these wild foxes.

I presented a formal description on the behavior of these foxes, and, during the question-and-answer part, I had a sense that their fears had largely been replaced by curiosity.

"Never had a gray fox sleep on my car," one woman said in a sort of dreamstate.

"What did you feel?" I asked.

She paused and looked directly at me. "Feel? At first I was scared. Wouldn't you be? That was a wild animal. I had no idea what it might do."

"Were you afraid of the fox, or of uncertainty?"

"I don't know, but I remembered that I had to pick up my daughter from preschool, so I made a noise."

"What happened?"

"That fox woke up, sat up, saw me, and jumped off the other side of my car."

"How did you feel?"

"Relieved."

"What if I told you that the fox was named 'Sleepy'?"

"You mean it's a tame fox?"

"No. Still wild, but he loves sleeping on cars, so he's come to be known as 'Sleepy.'"

The woman asked, "What should I have done? That fox was lying there on my car all curled up like *this*, and

I had to pick up my kids. Nothing like that had ever happened before. I was afraid. I didn't know what to do."

I asked, "What noise did you make?"

"I had these papers in my hand, see? I was so afraid, so I shook them, or they shook and made a noise. I don't know."

"And?"

"And I don't know where he went. Could be anywhere."

"These foxes are both audacious and timid. You could have just opened your car door and gotten in, and the fox would have jumped off and disappeared. They won't bite unless they have rabies, and these urban foxes around here do not have rabies. There hasn't been a reported case in over fifty years."

A few other questions came, but then one of the managers asked, "Can we have a list of all of the Fox Guy's recommendations for dealing with gray foxes?"

The lead manager shot back, "Oh, yeah, I'll have them made up and send a copy to each department. Pass 'em on to your people."

Weeks after that talk, I spoke with someone who had recently been hired by one of the high-tech start-ups, and she told me that the hiring manager had said,

"Don't be surprised if you go out one afternoon to leave and there's a gray fox sleeping on your car. Happens a lot around here. By the way, the fox's name is Sleepy."

Why are so many driven by fear, especially of the unknown? The concept of wildlife, or "wild," carries so many threatening connotations and implications, so many fright-filled illusions for us—stuff we don't consciously understand, rooted in the depths of our unconscious. And it was all made up by us. We call it reality, but isn't this all a fabrication, an accumulation of stories, even just fragments of stories we tell ourselves and others? Feels to me like it's all fiction. Isn't it time that we tell new, fresh stories about our fellow creatures, and thereby reshape our relationship with the wild? Shouldn't it be given a new name?

TWENTY-THREE

THE WILD ONE

Mama Bold and Gray still associate with one another even after all that they've lived through. They have been together for a good long while—possibly longer than most foxes. Maybe time has shown them how to recover when nature takes its toll.

I hadn't seen Mama Bold or Gray for at least six months, so I went looking for them and found myself over in the eucalyptus grove next to Skunk Hill, and there I called the foxes. None of them appeared, and so, remembering that Gray liked to sleep beneath the trees, I ducked in, but he wasn't there. I came back from under the dirty, spider-webbed canopy, and just there, on the other side of the chain-link fence, coming through the crisp, dry grass, trotted a fox. Suddenly, it ran to the passage beneath the fence nearby, ducked under, and trotted toward me. I realized that it was Mama Bold coming to say hello.

After all that time, she still knew my call. At just about the same time, her mate, Gray, slipped under the fence and came up the wildlife trail. They met no more than five feet from me. They touched noses, but there was no tail-swishing like there had been when they were young. They'd been together for far too long for that display. Bold looked tired. She ambled along the trail.

It will be interesting to see how she does next year when she and Gray have a new litter. Will the weight be lifted? Will she run again and chase Gray from November through February? Will she flop onto her back when he comes up to nuzzle her? How will she respond to her new pups? She will know and feel those chains called memory, those chains that so freely bind her to her own survival. Life remains, persisting yet raising perpetual issues for all wildlife that forage in the starlight.

I knew that an ending was coming when Squat and Bold (not yet a mama) stopped hunting together. That morning, he came up on the hill from Fox Hollow. Bold, his nearly mature daughter, sank beneath the Italian buckthorn bush, avoiding Squat's sharp stare and pointed ears. The tension between them was as a taught bass string, ready to twang. Just watching, I felt it. He went on up over the hill and she vanished back in the brush. That was it. From then on, they went their separate ways.

Squat's mate, the "Wild One," as I called her, must have cried caution and shouted, "Goddamn this crazy insanity in the background! How can you even begin to trust those human beings? How can you walk out there like that? Don't you care about your life? Don't you care about the little ones?" She was stressed from a thousand vexations coming at her like mosquitoes. She'd not known a so-called urban home region like this. She had come down off the range up there on Jasper Ridge, near the reservoir. She was not used to seeing human beings, and I seldom saw her except on my trail cameras as she groomed her pups. She was an attentive mom, but like most, she had her favorites.

Always aware, she understood the threat of anything approaching. She backed off into the brush. She watched

169

as it passed, bobbing her head from side to side, up and down, seemingly measuring the intruder beyond her immediate sight. She was afraid that it might smell her. It passed on. She sucked in its scent. Identification established, she relaxed for a brief moment.

TWENTY-FOUR

LET IT BE

Early on, Squat and Bold trotted out along the back road hunting. They'd bound into the weeds and come up with a smoky gray field mouse, then tease it until the mouse was crushed or frightened to death. Bold would then pop it down as though it was crunchy candy while Squat watched. I stood there, also watching through my camera's lens, photographing them. Such feeding continued into late February. They were father and daughter, bound together because she'd been born in his natal den, and not because she carried any of his genes.

I'd call to Bold in the early morning, and when she was nearby, she would come from the brush out onto the old dirt road. She knew I was there before I knew she was there. She'd walk right out from the brush. Sometimes she hung around near her natal den, but at other times she was on Fox Hollow Hill Trail, or around back at the water purification plant, or there in the weeds, the alkali saltbush, and the eucalyptus.

The area around Skunk Hill was a supermarket of mice for the gray fox. All foxes hunted there. The year before, Bold's wild mother had taken her there. Back then, Bold was but a pup with a pudgy nose and hardly any color at all. Her birth mother was the Wild One, the one that had followed Squat to the Baylands from the mountains. I could never imagine why she came down from the mountain.

Little Bold knew the eucalyptus grove and Skunk Hill, where she learned some good lessons. When she was young, chasing her siblings, playing, dealing with food issues, and learning to become a full-time adult gray fox, she mixed her experience and her knowledge with the environment: the trees, the water, the flowers, the birds, the food, and the insects that sometimes caused fire on her nose. It all became a medley of life. Somehow, she understood that she would live in that environment, in Fox Hollow, for the rest of her life. As such, she needed to learn its dynamics, and that included its people.

Sometimes she didn't show up in Fox Hollow

because the night before, prey had eluded her, and so still she roamed out there, hunting late into the warming morning, hungry. But then there were days where she'd hunt eat field mice like popcorn out along the back road, ears twitching, picking up the slightest rustle of grass from those tiny mice feet in the weeds. She'd learned it all. She understood it all. She poised, waiting until that exact instant to launch into a perfect arc, then come back from the brush with a gray mouse in her jaws. How did she snatch them up from off the ground without injuring herself with a snagged tooth or twisted paw?

Sometimes she wanted to show off by what many call "teasing" her prey. The lengthy ritual went on for more than ten minutes: she'd toss it high in the air, catch it in her jaws as it fell, then toss it again and again. The younger foxes tended to play with their food like that, practicing. Their elders already knew how to quickly kill and move on to the next. They no longer needed practice to be exact.

That practice of tossing a catch, sending ungodly fear deep into its heart, is not the practice of true teasing. In my eyes, we have clearly mislabeled this activity. Let's stop calling it "teasing." It is more like "compulsion," a need to know that its prey will not run away. Cats do it and so do young gray foxes. Maybe by renaming the action we can grow closer to understanding how death functions in the wild.

It feels like a hardwired behavior to me, but one that requires practice. There were times when I nearly intervened and cut off the "teasing," but I stopped, reminding myself that this drama before me, this fox creating such fear in that mouse was the very engine of nature at work, shaping and molding the environment and everything that lived and existed within it. I told myself, *Look at it. See how it works, but let it be.*

I saw the aftereffects, too, when Gray brought home meat for his pups. That's when the hierarchy in the gray fox community is most clear and honest: when food is at stake. The adults have already established a hierarchy for their pups. One is the alpha and one is the omega; the rest are supporting roles.

I saw the hierarchy in action one afternoon. Gray had been out hunting. The pups were all gathered there on Fox Hollow Hill when Gray came up the trail, brushed past the group, and dropped the squirrel before the alpha pup. She ripped into it, peeling off its fur and sinking her teeth into the bloody flesh. She dined alone, chosen by Gray.

The remaining four pups watched. Not all had learned the rigid hierarchy. Pup number two in the chain dashed in and snatched away a thigh. Instantly, the alpha pup attacked her sibling, chasing him, shrieking, barking. Lesson delivered, the alpha took the remains of the squirrel back beneath a bush and feasted.

Gray might have said something like, "We ate almost everything we hunted, and even what unintentionally crossed our paths. When I knew that my pups were hungry, I went out and brought home a young Canada goose. Had to stop, drop it, pant, just catch my breath, you know? Then pick it back up and trot on till I could dump it at Daring's paws. Yeah, I know it was big, but my pups hadn't had a decent meal in the past four days. Had to have a big enough bird."

TWENTY-FIVE

SQUAT VS. BOLD — THE FINALE

That time I saw Squat and Bold together, her rack of forty-two teeth was flaring from her gaping jaws. The ripping, the biting, the shrieks, and those teeth filled all in a terrifyingly blurred fury of six seconds. A long fight for any fox. Time and space froze at that moment. I felt a physical pause. Squat and Bold pointed past me, out toward the marsh. I half-turned, looked back over my shoulder, and there sat a fox, watching. I'd later call him Gray, the master hunter. Bold ran to him and they have been together ever since.

One morning, less than a month later, I came down onto Fox Hollow Trail with, as usual, Bold following me. She'd already followed me for some two hundred yards—a habit for her.

Ahead, on the opposite side of the chain-link fence, Squat came through the tall grass, moving at an easy gait. Maybe five minutes before, I had called his name, and now he came.

It was just then that Bold came up to where I stood, and in that instant, Squat saw her. Everything was electric. He dashed along the fence line. Bold shot off along Fox Hollow Trail as he slipped under the chain-link fence. Bold disappeared over the edge of the hill and onto the dirt road.

Squat appeared alongside me, his nose swishing left to right, sweeping the grass, and in an instant he caught Bold's scent and dashed off, down into Fox Hollow.

Stunned, I stood there. Why was Squat chasing Bold?

I had no idea what had just shook me down there in the pit of my gut. It'd happened in mere nanoseconds. For a moment I thought, "Well, that's it." I mentally shrugged. "They're gone." It was something within the lives of these foxes that I'd never seen, yet at the same time, I somehow knew that whatever it was, it was critical. Something was about to happen.

In spite of it all, I jogged along Fox Hollow Trail, and when I came to the edge where it goes down to the dirt road, I saw that Squat stood motionless, looking up and across to where Bold stood, her tail arched in that severe arc with her humped back that cried, "Beware. I'll fight, damn it. I'll fight. Back off!" It was the way a gray fox gave notice before either knew sharp teeth and tight jaws.

I knew the signs, but this standoff? Daughter against father. I quickened my pace, then stopped, breaking for pictures once I reached where those two foxes were squaring off. All froze, silent. Tension bristled. I stood stunned.

Bold stood frozen up to my left. Within that instant, she charged down the road toward her father, shattering the stillness. She put herself between Squat and that legendary natal den off the road where she had been born, that den where she wanted to give birth to her own litter. She wanted her turn.

Squat needed to chase his daughter away, to defend that den, for he had been born there as well; he too had fought for that natal den and won it from his parents. At that moment, Squat repeated his father's history. I wondered if he was aware of the juxtaposition, if he had seen this day coming, if he had prepared for it.

I didn't have time to lift my camera, position, and shoot. I thrust my camera in her direction just as she crouched, bared her teeth—my camera flashed—and they slammed together in a twisting, churning frenzy of white bellies and red ears. There was some growling, but they moved so fast that I couldn't take it all in until suddenly, time seemed to reset itself. Both foxes looked beyond me, out toward the marsh. I turned. There, watching the fight, stood a gray fox. The moment that Bold saw him, she ran to him, and the two of them moved off along the edge of the marsh.

I stood there looking out across the marsh, my mind elsewhere as confusion boiled through. I tried to understand what had just happened, but in many ways it was alien to me. The fight was too fresh, too emotional, too complex for me to think about.

I couldn't track Bold or Gray, as I was to name him later.

I turned back to where I'd last seen Squat, but he wasn't there. I looked around. He wasn't there. I called. He didn't come. That was the last time I ever saw that seminal gray fox who taught me what it meant to be *Urocyon cinereoargenteus*. That moment ended a saga. It was the end of an era. Squat vanished.

Bold had fought for that ancient natal den and won. She had a mate and needed a place for their babies. She birthed her pups—and earned her title as Mama Bold—where she'd been born, and where her father had been born, and where untold generations of gray foxes for more than twenty-four years had been born. In fox-time, as such a thing surely exists, the den was ancient and worthy of preservation.

I came to call that male fox "Gray" because his whole face bristled a light gray, instead of the usual rusty color around the ears. He was unique. His muzzle was just a bit too long for his face, and he passed that down through generations. In profile, he sported a rather flat face from his forehead to the tip of his shiny nose. Gray was skinny and tough. He was an exquisite hunter, a provider for the den when Mama Bold was off duty. She'd already watched him hunt. That's why they'd paired up, and that's the way it's done, at least from what I've seen. She decides when they will be a pair and when they will not. Their strongest bond, I think, was that they enjoyed each other's company.

Given enough territory, the yearlings disperse. They vie for a mate and, once attained, once she is pregnant and she has accepted him, then territory is established, marked by feces in the middle of the trail or some other conspicuous place where all can smell and see and know where the lines are drawn. In late November, at least along the edge of the San Francisco Bay, both sexes pursue the other and develop monogamous relationships. They welcome each other. As time draws near to her estrus, the male sniffs the base of her tail, assessing when she will be ready to accept him fully and possibly—but not necessarily—forever. She is aware of it all, and her body responds in kind.

In my eyes, Squat knew that Bold should have dispersed long before, but he tolerated her until, as the father of the litter, he had to demand that she leave Fox Hollow. That's the way of the gray fox. He tried to tell her, to make it clear she had to go, when he stopped hunting with her. According to Squat, Bold had clearly overstayed her welcome. According to Squat, the way he saw it, she needed to go find herself a mate, set out on her own, and there begin the cycle all over again. To his mind, it was imperative that she leave Fox Hollow. The species had to go on through her, but elsewhere.

TWENTY-SIX

PROTOCOL

It was that time of year in late fall, early winter when there were no territories and no boundaries that needed staunch defenses because there were no gray fox pups to teach, protect, and feed. About that same time of the year, rustling trees set leaves free to cover the ground. The pups had, for the most part, dispersed, leaving their home range in search of mates and new homes that they might scratch from the meager landscape along the San Francisco Bay.

Sometimes their areas overlapped with the others nearby. They came to understand that there was not enough land, not enough cover, not enough food or water, so they kept moving.

Somehow, they succeeded. Somehow, they found an acceptable mate; somehow, they established a natal den of their own or stumbled upon a vacant one. After that, they needed to understand the landscape, find out how much food was available and for how long it may be sustained, for they knew that their pups would need protein in increasing amounts. That meant that they would eventually need to extend their boundaries.

Their need for protein had much to do with the development of the myelin sheath and the neurophysiology of their brain. Whether they thought about it that way or simply felt the instinctual impulse made little difference. All that mattered in the end was having the right food at the right time.

They went about marking their territory with urine, with feces, and with gland secretions placed strategically, so as to give warning that beyond the gate along the back road, that land was claimed.

Brownie and Little One, along with Helper, controlled that territory, and their domain was spread far and wide. The scat and urine message was clear: "Beware to any gray fox that dares to cross this border. We will fight for our family's space." The coming generation is that important. Protect the territory, because the pups depend on adequate food coming from the land, carried in the jaws of their parents. Foxes heeded the warnings, but only sometimes. Screaming fights exploded through the night air whenever their territory was breached.

At the Baylands along the San Francisco Bay, it didn't always happen like that. There were some foxes whose foxsinalities were unsuited to be parents, who nevertheless gave birth. Sometimes the pups suffered from neglect.

So I asked myself, "How did these newcomers, these young dispersing gray foxes, know how much territory they needed to stake out so that they could adequately feed their own? Did they sample a portion of the landscape to get an idea of how many rodents and other such edibles populated the region, then do a few mental calculations based on square acreage and prey population? Did they extrapolate essential data? For the second-year parents, they just knew how much territory they needed to keep their pups and themselves alive, but who is to say they didn't apply some methodology?"

Brownie, up there on the hill, displayed mastery of understanding and defending his landscape as needed. The pups successfully hunted at just shy of two months. By then they already understood the protocol.

Around the end of November and the first part of December, in most cases, the young foxes leave their home range. They disperse much like late teenagers tend to leave home and go out on their own, to explore and move beyond. That's the cultural expectation for foxes as much as it is for humans.

Dispersal is the time for foxes, male and female alike, to strut. Giving and receiving attention is the name of the game. Males watch the females, and the females watch back. They vie with each other for territory and suitable mates. They frolick in polygamous encounters, playful and vivacious. Everyone is fair game, and if one young fox doesn't play the game right, the chiding and heckling from the others is frightening. It is a time for raucousness, for testing, for making acquaintances.

Then that time passes, and somehow they almost always end up in pairs, drawn to one another in firmly monogamous relationships. The wild oats have been sown and it's time to settle down.

TWENTY-SEVEN

HELPER

At certain times of the year, there are patterns of items out along the trails crossing through fox-land. To the average person, the walker, the jogger, the runner, and all in between, these items are invisible, or if seen, people assume that the feces along the way must be because derelict people don't pick up after their dogs.

For one who sees, hears, smells, and considers the environment, scat is filled with knowledge. From March through mid-October, when territories are defended and invaded, gray foxes speak with scat, urine, and sometimes their scent glands. It's a kind of signage that might say something like "No Trespassing" or "You'll Find My Teeth in Your Thigh." Threats are assumed, but the true message could mean just the opposite: "This territory is set aside to feed the young foxes that are up the hill playing and chasing in the den area. Do not cross this border for their sake."

For most, that warning was enough. Some others who came to these territory markers ignored the warnings. When they met the fox that left those indicators, they seemed not to know that a hiked tail meant, "Beware, I will fight. I can be aggressive if I need to. I'll give you room to leave, but make your choice." With their bushy tail up, back hunched, legs extended, looking huge and striding around in circles, the fox gave an overt message of defiance, anger, and protection. Some foxes intuitively avoided trespassing. Others not so.

The territories on the north side of the Madera Creek were firmly delineated, wholly defended for whatever fox reason, until, from out of the marshes, from out of the brush, came Helper, slow and easy, smooth as she walked. She was the very first gray fox documented as a helper female, thus earning her the name.

Helper came from the brush and up onto the levee road at Madera Creek. I remember. I was down the road taking pictures of Creek's family and all the happenings there. She popped out beside the road and came down in my direction as if she knew. I thought, *She's invaded the territory and will surely be chased away*. But she came as if she knew everyone. When she arrived, it was obvious that she was known by all of the foxes in the area as they came from the brush, tails swishing with happiness, the pups displaying their place in the hierarchy.

As a helper female, she foraged for the pups; she took them out to find beetles, to find the right kind of grass to make that old tummy feel good again.

While all of the other foxes would trot alongside me to keep up, from the beginning, Helper lagged behind. She sniffed the roadside licorice and tasted fennel from the time that it was but a foot high, ate grass that grew along down in the creek's bank, lingered, looked, and listened. She seemingly enjoyed just being. She had no need to keep up until I turned and called, "Come on, Helper. Come on," and she'd pick up her pace a bit. Then, in the dark of the morning with my headlights hitting her eyes so unaccustomed to such a glare, I called, "Come on. Come on, Helper. Let's go." I thought I might need to repeat it, but then she turned that walk into a quick trot and followed more closely.

Helper was the only gray fox that I have ever known—and I've known a lot of them—that was a renegade among the gathering of gray foxes, unrelated but somehow connected by a bond of friendship. She was accepted regardless of her outsider status, and such a thing is nearly unheard of.

Out along the levee road, along the perimeter, four gray foxes—Tippy, Tense, Helper, and Blue—held their territory although there were no pups involved, so there were no little mouths to feed. They had only themselves, and they were qualified hunters. That meant that in the big picture of gray fox culture, territory was not dependent upon a litter that needed meat brought into the den.

With no pups to feed, why did they still mark boundaries? Why did they still establish territory? Only part of it was governed by routine; after all, they too needed field mice and mallards, but with no litter to protect, why should the territory matter?

So many territorial battles had taken place across the years, but along the perimeter road, somehow peace governed.

Around the intersection where the back road hits the perimeter road, Brownie had marked back into Tippy and Tense's territory by about twenty feet or so. He was smart. I'd known that from the first time I'd laid eyes on him. He was a fast learner. He didn't always let me, the human being, lead the way. I admired him. He didn't want to leave the intersection as open territory. Technically, it was free, as is any area where two boundaries rub up against one another, but Brownie marked it so that he governed every inch of road. Brownie owned that territory. He was compelled to defend it.

Brownie saw the four foxes coming from Palm Corner as invaders, and they made him cringe. He saw the vision and laid out a plan to do battle. He knew it was coming; he just didn't know when.

Tense was leading the foxes down the road, away from Palm Corner. Tippy objected. She wouldn't go beyond the bench. Tense didn't care. She continued with Blue and Helper following. Ahead at the junction, Brownie appeared, tail hiked, back arched, standing tall on all fours. Bigger than life itself, he stood. To Tense, he was a formidable foe, yet she needed to confront him.

That's when I approached and saw the situation. It was dire, and yet, I didn't want to intervene. Given the circumstances I'd just stepped into, I needed to know their behavior. After all, wasn't I studying them, not just watching like some soap opera addict?

Brownie charged at Tense, and the two of them screamed and squealed like a power saw as they tumbled and bit at each other, white bellies flashing, ears bitten, and within seconds everything froze, motionless.

Helper looked at the absurdities playing out there in the dust. Brownie looked huge, his back arched, his tail shouting "threat," "danger," "battle." In an instant, she challenged him. She shrank down low and bared her teeth at him.

Brownie stared down at her, then looked to the others. He recognized Helper, the gray fox he'd been friends with for more than two years. Stunned at the sight of Helper, Brownie's mind stumbled. He didn't know what to do. He had planned to battle these invaders, these four foxes who had come down from the palm trees over there along the levee road. Helper, who had been Brownie's friend, the fox who had hunted with him, who had slept there in the grass nearby, was one of those from the palms.

Confusion waffled through mind and body. He wondered, "Why is my friend with these trespassers?"

Helper instantly disarmed him.

She was there beneath him, lips rolled back, her teeth glaring up. Resolution swam through her. Though masterfully intelligent, Brownie failed to understand.

Brownie asked, "But here Helper defends these renegades. Why?" Confused and disoriented, he first went up the road toward the old observation deck. The next moment he turned and trotted back to the junction and up the back road maybe thirty yards or so, with his tail hiked and all of the other foxes watching from the other side of the old bench.

Helper intercepted him, laid back, bore her teeth in a grimace, and "announced," in no uncertain terms, that he would surrender. With no other options at hand, he stood down. Prior to this, Brownie had claimed the entire road on the top of the hill along the back road and on to East Bayshore, a long strip that extended to the Saltwater Wetlands, but now he had four friends to join him.

TWENTY-EIGHT

THE THREAT OF BLUE

Each time Dark came down into the big clearing and Blue was there resting, Dark charged him and Blue would flee through the brush, plunging through the thicket since there was no time whatsoever to take a trail that twisted toward escape. Trails are obsolete with an alpha on your heels.

He didn't know what to do. His mate was already with their squirming little ones inside when his world came thundering down, an avalanche of forces. Dark could not have Blue anywhere in the vicinity because he presented possible competition over the females. Blue was also a drain on food supplies that needed to go to the pups. Dark knew some of Blue's trails, some of the same paths he'd take night after night. They all had them. They all led to either food of one kind or another, or to a clearing where they could relax.

Dark waited along the trail back in the thicket. He remembered the fight that he and Blue had had earlier that morning. He felt that he could have pushed him even further. This one, the one right now, must be decisive. This one had to be final, or so Dark sensed. It was time to send a sterner message.

I heard the pained screams, the hurt yips, defiant barking back and forth, ripping into the other fox's hind leg, a tooth puncturing the base of an ear, and the ear torn free to bleed in the dirt.

The adversaries parted and walked away. Blue did not want to be further humiliated, nor abused by Dark. He went to the north side of Madera Creek where he hung out with Helper and sisters Tippy and Tense.

Meanwhile, across the land, the males sniffed beneath the tails of every female; even fathers sniffed to see if their daughters were close to estrus. The males were aroused and kept track of the females' states of readiness. Sometimes they had erections for days.

TWENTY-NINE

FINDING BONES

Early February, Gray showed up again. I was happy to see him come trotting down into Fox Hollow with Mama Bold nearby. I knew that they made good mates, although there were times when Gray would be gone for more than a month and I thought for sure that he would stay gone, but he always returned. He hung out with Mama Bold, and just liked being with her. She knew it, too, and regarded Gray with affection. They became familiars. I saw several times when Mama Bold simply melted when Gray came around. One morning in the dark, down on the road, they came together, he from one direction and she from the other. When she saw him coming, she rolled over onto her back, all four legs pointed to the sky, and he nuzzled down on her belly. It was easy to know that these two wild foxes cared for each other. They bonded. This time it would be for a good long while.

That year, once Bold had taken the den, she and Gray were to have pups. I kept track. I expected to see them, see some indication that her litter had been born and was growing up, but there was nothing.

It was mid-April and there were pups. Because there were so many, I focused my attention along Madera Creek and seldom saw the foxes in Fox Hollow.

But one morning, I decided to go down into Fox Hollow to explore and find out why Mama Bold had no pups.

I went over to Skunk Hill and explored, looking for something—not knowing what—over on the edge of Windy Hill where the foxes sometimes hung out. I trampled the grass. I looked across and saw that alkali saltbush right on the edge of the road. There seemed to always be dead things there, like half of a baby skunk, or the thigh bone of a squirrel, or the remains of a swallow. It was like a dumping ground for unwanted parts.

Just under the lip of that bush, I spied a skull lying within the broken remains of a rib cage. I picked it up, and, after counting the teeth, I knew that it belonged to a newborn fox. The skull of that pup, the little one that should have burgeoned into a magnificent gray fox of the Baylands, lay in my hand. Mama Bold had squeezed that pup from her body, had known its face and smelled its scent. What force had carried it here, to this place of refuse?

I held bones.

Judging by the size of that skull, so tiny in the palm of my hand, it'd had only enough time in its brief life to know the presence of its mother and nothing more. In a just world, it would have been born and nurtured, then opened its eyes on the world after one full week of quiet and love. In another couple of days, it would have begun to hear and grow accustomed to the utter cacophony of the world. It would have known the texture of Mama Bold's tongue as she'd groomed it with care. Whenever it was hungry, it could have found its mother's nipples and there suckle her warm milk, safe and sound, and it would have grown and grown and grown to fulfill the measure of its existence.

In a just world, we would not hold the bones of our little ones.

Red light roamed across the landfill in the chill air. I stood and questioned what in that entire region would be predisposed to kill that pup—or worse, the litter.

There were no predators nearby, no coyotes, no pumas; there were bobcats and golden eagles on rare occasions, but very seldom did they cross paths with a fox.

I needed to find out how that baby fox had died, and if there were any others.

Beetles fell from that skull. It was dirty and raw. The cause of its death was the reason why I'd not seen any pups in Fox Hollow that season. I nestled the skull into a plastic bag that had once held my lunch, and took it home.

THIRTY

ASK THE UNIVERSE

I have been going through a period in my life where I need to examine the fringes, the subterranean depths, of who we are as human beings. Time grows shorter every moment. I stumble in the darkness for answers, but lately from a much deeper, more important foundation. This has all grown from my engagement with the foxes, those who travel in what we see and understand as the dark of night, contrasted sharply by those who dwell in the daylight, those who run roughshod over everything sacred, including the Earth.

These predators from the darkness present me with powerful questions: Why aren't they more like me? When is the right time to act? How can we come to know that the truth lies in perception?

Something primordial rumbles, "Act when you feel. Therein lies self and truth." These gray foxes are only a sampling of archetypes for their kind. Within a spectrum of being, they have ever been revered or demonized throughout history. Some have elevated them to the status of gods. The ancient Greek slave, Aesop, used the fox to teach us lessons and morality, and such memes persist even through the madness of modernity.

I ask the universe, "Do gray foxes communicate?" and am met with silence. No answers descend from on high, no voices echo my own. People are certainly not chatting about it over any of their social networks as they share and like back and forth, talking, purring into their cameras, congratulating and condemning each other's humanity. They do not seek to know more, to plumb the depths of their lineage and find the connection; they are content with the universe, or perhaps indifferent.

The raccoon cubs sometimes use my trail camera posts to climb. Then there are the raccoons' climbing trees: swamp mahogany trees back there in the thicket that they use to practice their agility. Even the young ones rarely fall.

And there are the skunks, the opossums, the woodrats, all paying strict attention to the others. They are interwoven.

Is that necessary in the world of the gray fox, this connection? If so, how do they transmit information to each other? In their world, is there something similar to what we call information? Maybe not. Then what? What if the answer is nothing? Maybe everything hinges on interpretation of communications that are entirely alien. Is it possible to understand their paradigm, an alien paradigm?

I want to find out. I need to know, yet I am largely alone.

THIRTY-ONE

A DIALOGUE WITH SQUAT

Squat: I think we can communicate like this because we understand.

Me: You're probably right. I've wanted this moment for years now.

Squat: We finally learned the framework of knowing each other.

Me: Which is?

Squat: The way I hear it, you don't use your senses like I do. Somehow, evolution has made us different.

Me: Not sure I understand.

<u>Squat:</u> Oh, sure you do. Had you not understood, would we be having this dialogue?

Me: Oh, yeah, now I remember. I think I was down in the channel one afternoon after being with the foxes back there. After years of watching them and documenting their behavior, I saw that you foxes use your ears far more than I could ever imagine. I wondered why.

<u>Squat:</u> And you arrived at the same conclusion that I did. Well done. You humans use your eyes far more than my kind. We are not the same, but so similar. You use imagery to understand and delight in your world, whereas we foxes hear and smell to build our mind-maps.

Me: Yes, yes. So true. And you don't see, you don't visualize your world; instead you hear it and that sound gives shape to reality. Well, no, not exactly; it's more like sound is reality minus everything visual. But I can't quite get that notion through my head.

<u>Squat:</u> You depend too much on your eyes. And see what that does to you? I know. You know. Now we bridge the gap between us and carry on with some understanding.

Me: Understanding could bring change. Or revolution?

I once noticed a pup lying in the grass, asleep. I watched. His hind leg twitched. His ears honed in on something going on in his dream, and I wondered whether his spasms were auditory and olfactory responses translated into understanding.

Squat: Exactly. Yes, that's what we do. I dream in soundscapes, tones, some smooth and at ease while others are sharp and strident, and together they sculpt meaning within our dreams. That is but half of our ability to know. The other half is in our nose when, as you say, we smell.

Me: So now you're going to complicate all of this, right? You already have auditory ears and an olfactory nose. Are you going to tell me you understand the world through a sound embedded in an odor, or an odor embedded in a sound? I don't know how to make sense of this.

Squat: Standing on a beach, facing the ocean in a storm, the dark waves wash high onto the sand. How many messages are contained therein? The higher the waves reach across the beach, the further back you move. You are reacting to the sound as much as the image of that wave, and the possible cleansing or exhilaration or death that it contains.

Me: Then what do I have to do? How can I learn to communicate with you, and you with me? I mean, how do I continue to know what you divulge? How does it extend beyond this singular conversation?

Squat: We have to learn how to understand each other first—our reasons for being. I have to learn how to refine my knowledge of your tone of voice, and sometimes your words get in the way. You have to—

Me: Yeah, I know. I have to let go of words. Well, maybe not completely, but become more flexible in my perception of them. I have to see through a different window in time.

Squat: And that's an immense undertaking, but it is possible. I want to introduce you to someone. Remember that morning? Remember when I introduced her...?

Me: I know. I named her Bold, and she became the matron of the region. After she'd had her pups, I called her Mama Bold. I cherished that name. She was such a special fox.

Squat: I knew that. That's why I brought her out from the brush that morning and introduced her. I knew that she would come to be a power in Fox Hollow. She chased me away because I let her. I knew that it was time for me to move on, because that female I paired up with last year was already gone. I couldn't give Bold the home range without a fight.

Me: You two gonna fight?

Squat: Of course. That's a given. And I'm gonna lose. That's a given because that's what we choose.

Me: How can you choose? If it's your choice, why fight? Do you mean to say that gray foxes know about free will?

Squat: I can't explain that, not now. Not yet. But keep trying. Keep asking and watching, and we will show you in time.

Me: I will. I'm not going anywhere.

Squat: Glad to go exploring with you.

Me: Glad to go exploring with you.

THIRTY-TWO

TAIL TALK

I think I have unlocked certain parts of their "vocabulary." For instance, the most dramatic expression is when a fox arches its tail. It looks like the start of a knot, an inverted U right where the tail emerges from the body. Often, this action is paired with extending their legs, making them look far bigger than they truly are, intimidating with an arched back and a ferocious pose.

At the same time, there are variations in behavior from one fox to another. Sometimes the arched tail happens just before the fox dashes off through the dense brush to do battle with that trespasser. Other times, the arched tail leads to a stutter step and playful hop. Still others, the tail unwinds, and all becomes calm again.

Why is it that every time two foxes meet, even if they are mates, there's always that hesitation, that stare, sometimes that hunched back and arched tail, marching toward the other, only to smell-see and recognize? When the alpha female, my Buddha dog, Dark Eyes is in the big clearing, Blue enters from the Ivy Trail, and when he smells her, he goes to her in that usual submissive gesture and says hello. And in the process, she is known.

Brownie is famous for doing the tail arch, but just before he goes on the chase, that little male does a dance, almost ballet-like, prancing about as part of his warning. Like a bullet, he shoots after the trespasser, most often Tense.

At the opposite end of that tail, the fox is paying careful attention to every twitch of movement, every whisper of sound, for clues on how to proceed, and the reaction comes instantaneously.

It's not until the foxes physically engage that they make any sound whatsoever, be it a fight, a chase, or a moment of pleasure. The urban foxes that I study are, on the whole, remarkably silent. When a gray fox shudders with pleasure—and that may come in many forms—there's always that squeaking, that yipping, and vibrations of pleasure seem to consume its form.

It is unmistakable. Those sounds, so rare and poignant, speak volumes. There is an interplay of sharp, high-pitched squeals and yelps of enjoyment that caresses my ears on occasion, but overall, the gray fox is not prone to making sounds—at least not the foxes that I study. On a day-to-day, hour-to-hour basis, these gray foxes are unerringly quiet.

The twitch of a gray fox's tail sends a coherent message to all present. Sometimes a gray fox will swish its tail, signifying pleasure, happiness, or thrill. A tail hiked holds meaning. One ear flopped back is a mystery to me. Is that their language? I see it and interpret that particular behavior, but at the same time I have to ask, "Is my interpretation accurate? Does it say anything about direct communication and how that's accomplished?" I can't tell for sure, but to some extent, yes, for once its tail hikes and its ears are thrown back for protection, a fight shrieks as they tangle, bite, and cry out. That stance announces a course of events, repeated again and again before me.

Does that swishing tail always mean the same thing? I don't know, but from what I have seen, it presents happiness tempered with a dose of deference. There is often swishing when a pup approaches an adult, which suggests that they recognize and enjoy seeing their elders. There is often swishing when an adult meets one of the alpha foxes in the area. A tail swish means so many things and presents a multilayered aspect to their communication. All of it deserves further study.

THIRTY-THREE

THE ROSE DEN

We sat there, Greg and I, and watched on the trail cameras each cold morning as the young dispersed over the landscape, unlikely to find a place of their own, much less a suitable mate, along the edge of the San Francisco Bay. For many, they found that in order to inhabit any part of the terrain, they had to share with others, and that concept was alien to the wild gray foxes.

They had yet to be massaged into the urban world.

From the time that Tippy and Tense had been born, the sisters were of the Southside Clan. Three gray fox pairs carved out a patch back in the thicket to call their own, each with their own den. Dark and Cute took the Rose Den, a thicket of rose bushes with pink flowers dappled over the outside; later in the year, fat rose hips would round out the season. The foxes didn't eat rose hips very often, but they devoured Italian buckthorn berries. Their scat was purple that time of year, filled with seedlings that birds would pick up and take away, to distribute as nature intended.

The gray fox has tools that it uses to grasp its world, to mold itself into meaning, just as we do. Because of that connection, we humans need to be aware that the gray fox's experiences have been developed in a universe outside of our own. They've felt the seasons shift, and the wet tongue stroking. They were born in the thicket, back there under that wild rose bush, a place that you or I could never consider crawling into. They have no problems finding their way. No, the Rose Den was well placed. It fronted the big marsh, with trails laid out way back to the thicket. It was home.

By late March, Cute was pregnant. Her breasts showed, bursting with milk. Dark hung around the Rose Den site every so often, but I had the impression that he was not cut out for fatherhood. Nor was Cute cut out to be a good gray fox mother. They were equally ill-equipped for the roles that they'd been thrust into. I only learned that as I followed the developments at the Rose Den, as Big Guy somehow managed to grow into gray fox maturity despite his inept parents.

The family survived, but things took a turn when it became seemingly impossible to find enough food, enough fruits, enough of the landscape to sustain health. A dark virus lurked just beneath the surface, and water became scarce. A sudden drought had swallowed their land and, unseen, the rodents scurried in the grass, carrying the taint of poison.

THIRTY-FOUR

THE WATERHOLE

Down there in Fox Hollow, it seemed as if the land itself had somehow started creating water. It oozed from the slope, and no one knew its source. I came down the slope from the backside of the water purification plant, looking for the cause of the newfound water, but I wasn't alone; the plant boss had sent out a crew to find the leak.

The chemists sampled the water. "Slightly saline," they announced.

The crew dug trenches, dug holes, and watched them fill. The engineers had no idea where the water came from. It seemed as if the whole hillside bled water. They installed pumps, but they could not keep up with the water's flow as the land wept, for none of the city engineers had an answer.

This became the only fresh water anywhere in the region. Summer stood shrouded in heat as the drought took hold. The crew had failed, and so they abandoned the job and let it lie. The foxes, raccoons, opossums, birds, wasps, bees, and skunks all came to the waterhole to the sound of the frogs croaking their elaborate hymns. It was their only lifeline, and they all had to share or perish.

The only other water around gushed from a pipe that fed the channel out into the Saltwater Wetlands, a saltwater marsh. The flow came directly from the San Francisco Bay, and saltwater swam through a two-foot diameter pipe and into their habitat. The wildlife there drank saltwater.

The waterhole at the purification plant had been there, full and filling daily, for about three months. As it did, the hole topped and overflowed, creating an unnatural freshwater marsh. Tulles grew up and into cattails. Animal trails from off the hill led directly to the waterhole. The wildlife hurried to drink their fill, somehow aware that the bounty couldn't last.

This all happened right in the middle of one of the greatest droughts in California's history. We felt its impact across the state.

One morning, I noticed that the water level had dropped, and from then on, it seemed as if the water was being sucked into the ground. I talked with people who were in charge, people who worked for departments in the city, people on the city council, and no one, not one, lifted their signature, lifted their voice, lifted a protest to help get fresh water to the wildlife. Their answer was to turn a blind eye to the problem.

The waterhole became a patch of drying, dark mud and leftover tracks from raccoons that had last come through when there was enough wet to lay down a track. Now their paws scrabbled across hardened earth, unable to leave their imprints as they left in droves.

Months passed.

I found myself out along the levee road while thinking of the waterhole, no foxes anywhere, when I hit upon an idea: a man I knew named Bernardo drove the water truck for the construction crew up on the landfill. He kept the dust down on the hill so that it didn't blow into the plant, and that kept him busy until he had to reload the truck. The drive back took him straight through Fox Hollow, directly past the waterhole, up out of the hollow, and over to the water purification plant to fill his tanks. On his way back, Bernardo passed directly by where I stood.

"Stop, hey, stop!" I hollered, waving my arms. He complied and rolled down his window to squint at me.

"What do you need, Bill?" he asked, concerned.

"Bernardo, when you come back down there by the waterhole, stop a moment and shoot some water into it. Fill it up, please. For the animals? You'll only need to do it once a week, maybe twice at the most. Can you do that?"

"Sure, but I might need to check with the boss. I'll ask him."

Bernardo did just that, but the boss told him, "No way. You just drive on by and get your ass up there on the hill. We got work to do and no time for those fucking foxes. Tell that guy to mind his own business."

I watched a few animals come to the emptied waterhole and suck moisture from the crusted mud until the earth cracked and peeled away. The water truck, heavily laden, drove by several times a day, and my friend waved his apology with each pass.

My only thought was *Water, water, water. The animals need water.* I needed to get water out there for the wildlife, for the foxes. I kept pushing. I contacted people within the city government, but no one answered, or if they did, they expressed concern but refused to act; they had never before dealt with anything like this.

I pushed more. I went to the plant boss's office and pleaded with him in earnest. At first he was irritated and told me to leave, but then something changed and he said, "All right, what do you want?"

"I want that waterhole filled," I told him. "When it goes down, shoot more water into that hole so the animals can drink. That's all I want, and you can make it happen. Give the word. Let Bernardo save some lives instead of hosing it all down a dirt road."

The next day when I came down the hill, I saw that the waterhole was full to the brim. I went looking for Bernardo, to thank him for what he'd done, but he hadn't come in to work that day. Someone new drove the truck, but they kept the waterhole full all the same.

I never saw Bernardo again. I suspect he was fired for his involvement with me, which I can't prove, but I wish him all the best.

The plant boss met me by the waterhole, and I thanked him. He told me his name was Frank.

"Do you think…?" said Frank, pausing to glance around. "You think if we put a tub out here, they'd come drink?"

"Sure." I watched him toe the dirt with a work boot. He took a deep breath and waved his hand.

"We'll put it right out there by the water pipe," he

said.

"That's a good idea," I told him. "And I'll put a camera on it to see what wildlife comes by. We'd probably see some raccoons, foxes, sometimes a skunk or two, maybe an opossum or a jackrabbit sniffing around. They'd come for sure."

Frank nodded and headed back to his office.

Today the waterhole is dry, but the tub—well, I fill that most every day. And every day, the animals come to drink.

THIRTY-FIVE

FOX-KISS

What is the purpose of the fox-kiss? The young fox approaches an adult, always coming up under the adult's chin or alongside its jaw to touch them with their wet, black nose, and sometimes a lick or two. This is to say, "Hello," and "I respect you as family. You are one to be honored and so I submit." Adults exchange fox-kisses too, but only under certain circumstances.

The fox-kiss is one universal behavior among foxes, or so I feel justified in claiming. I've seen it happen countless times, although there are slight variations from region to region, family to family. The kisses are one aspect of gray fox culture that lives through memes. The ritual is demanding, to the point that I have seen a pup pass by the alpha male (its "father" in many ways), miss the kiss on his first pass, then turn back to give his father a proper fox-kiss. Between the two of them, that sealed the moment. They had spoken.

Only once so far have I ever seen an adult or an alpha reject the fox-kiss.

Pup One Eye was born back in April with his brother, Sideburns. They got sick. One Eye held hope, but that was cut away when they euthanized him. 2016 was a year when tragedy thundered through fox-land.

One afternoon, Blue was out on the concrete of the overflow channel. Sideburns the pup ran toward him, to greet Blue and pay homage, too.

When Blue saw him, loaded with illness, eyes draining that pus and caking his eyes with blindness, he ran. Another time, he raised his head high so that there was no contact with Sideburns. Did Blue know that Sideburns' eye infection could carry over into his own eyes? Was that the reason why he didn't let Sideburns give him a fox-kiss? Do they understand that some illnesses are contagious? If so, isn't that a sign of awareness, of an elevated mind that deserves further investigation?

THIRTY-SIX

NECK RUBS

How much do their lives and their communication revolve around nonverbal gestures?

Neck rubbing is one of the most gracious of all their gestures, at least from what I've seen. Whenever the gray fox wishes to show great homage, it always gives a neck rub, sharing its scent with another in a kind of bond not found in humanity. It's most sacred, which is obvious no matter how many hundreds of times I've seen it. The neck rub is a gift among them.

Why then, with two foxes who clearly know one another, do they often give this elaborate greeting, even if they've only been apart for a few minutes? Generally, they greet each other as though it's the first time. The neck rub is like a hug with an old friend, but at the same time, much more than that. The ritual is freely given, even enthusiastic, each and every time, and the full meaning behind it has yet to be discovered.

Gray had a serious eye infection. He saw, but through a narrow slit into the world with his left eye. His right eye lay covered with a thick layer of crust, the dried drainage from an eye infection. He was essentially blind. Gray, his mate Mama Bold, and their tagalong year-old daughter, Pale, had settled down in the natal den up on the hill.

Ahead, there lay a broad open space of dead weeds. The hill flowed down to where we stood, looking up. Gray topped the hill. I called, "Gray, hey. Gray, how ya doing?" As Gray carefully walked down toward us, he saw Pale over near Mama Bold. I stood between them.

Gray went to his daughter. He went to her and he nuzzled her at first, but then he pressed his head against hers and shoved her head down to the earth, down into the dusty dirt. There he gave her a neck rub, sharing his scent with hers. His scent glands flowed at that moment, and when they parted, I stood there amazed at what I'd seen. It was clear to me, for when Gray approached, I saw that his eyes were clear and wet.

Pale had been scented by her father because she'd

licked the dried gunk covering his eyes that had nearly blinded him. She had licked his eyes free from the crust so that he could see. That's why he gave her a neck rub, showing his gratitude toward her for having licked his eyes clear once again. He'd given her his ultimate blessing.

THIRTY-SEVEN

SCAT

Tippy knows she's trespassing, knows that if Little One is down the road there, there's likely to be a fight. She goes anyway. Within the varied foxes, there is a point where one dislikes another and shows it. I have witnessed these dramas within the urban fox community. At the same time, there's a lot of cooperation, too.

When they strut around with tails hiked in an arch, back raised high, looking so much larger than reality, they sometimes perform a delicate prance on all fours, warning everyone that they are poised to defend themselves and defend their landscape as marked.

The smell along the road—the smell of feces—contains a structure of how the gray fox lives. The scent markings that you and I cannot smell or taste hold tremendous meaning within their community. Communication is instant. These are the invisible territorial markings, a form of language we do not possess. For instance, if the feces are left in the middle of the road, that clearly means that this area is taken by a specific gray fox. Such has been extrapolated from observing coyote behavior, and it applies to foxes, too. Often that scat is laid down on high places, even atop a flat rock, just because the mark will be obvious to others passing through. They must read the notice and understand.

Many do, but Tippy doesn't care because her reality is framed by the dimensions of life and death; in other words, she goes where she must to survive. Others trespass as a matter of course, as though they have no fear of punishment, or just don't care, or maybe no adults taught them about those territorial boundaries, those gray fox restrictions and protocols.

That arched tail comes with subtle messages, too, possibly saying, "I display this only because I have no other way to let you know I want no conflict." Nonetheless, when Little One charges at Tippy, growling, screaming like a banshee into the night, she shrieks, fur is torn, bodies trip and tumble over one another, and then they separate. That ends it. There may be injuries, but death is beyond anyone's designs.

The language of the fox is a mystery. An important part of it lies in the vast realms of silence. As I consider their language, I am often found wondering, "Is there no transfer of sounds? Is their language such that it has no exchange of information as we know it? Is the content of their language greater than our own? What would it be like to think as they do?

There are some who say that the paranormal fits in here. There have been interactions with the foxes that felt like magic; there is no other word for it in our vocabulary. A given fox cuts through the brush, only to meet me back in the big clearing—it knew where I would be and it knew how to get there through dense brush, while I had to remain on the trail. How? How did it know where I would be, where I would end up? Many with intimate connections to their chosen animals have stories to tell about when something happened that logically could never have happened, but did anyway. I try to will these experiences into being every time I am in the midst of seven or eight foxes all lying there, content, or biting at the vermin over there, scratching, depending on the time of year.

Daily I try to assess it, try to understand it. I wonder

why they seem to understand me far better than I understand them. Generations, millenia even, of human arrogance insist that I am the higher being, the greater intellect, the chosen ruler, yet each moment I'm with the foxes I learn how flawed that thinking is.

Just last week, early in the morning, the gray fox male Brownie and I came down the back road together. He's a fighter, and he suddenly let loose a challenge out of nowhere. Before dawn, in the dark, he charged down the road at another fox. The high beam perched on my hat displayed the action for my weaker eyes.

Brownie clearly announced himself as ready for combat, ears pasted flat back on his head, legs extended, back arched, and bushy tail flared and arched, as well. He pranced around on the road, showing off his wares. He was sheer elegance in motion.

The other fox stood no more than twenty feet away, ready as any good gray fox would be under these circumstances.

Brownie unleashed himself, charging at the trespasser.

I shouted, "Stop! No, Brownie. Stop."

He stopped. Why did he stop? Was it in the tone of my voice? But even so, why would my voice have communicated exactly what I'd wanted him to do? He did not have my language, yet he knew that something from my direction had insisted that he stop.

He turned halfway toward me and looked back.

I was stunned into inaction. Can I conclude that he understood me? Or is there something archetypal in such a cry that's understood across species of all kinds? Perhaps I'd simply startled him, but he hadn't run or turned his challenge in my direction. He'd just stopped and given me a questioning look.

Why can't I read them as clearly as they read me? I call Tippy and she turns; she comes back to me and looks up. She has two names, Tippy and Tip, and she knows that both apply only to her. Why? Why should a wild animal like her understand me? I call Tense, the female gray fox, and when I call, even if I'm on the north side of the creek, she comes. When we are face to face, she comes even closer and looks up at me. What does she understand about me? Why does she look at me like that?

If she were a dog, I'd reach down and pet her, give her a good scratch behind the ears or under the chin, but the foxes are not pets, and they are not domesticated; none of them allow me to touch them. So why approach to look with full attention at this human who offers nothing more than a familiar presence? Why respond to my call at all?

The questions eat away at me every waking moment, but oh, what magic I witness when my call is answered and an ancient being from the wild accepts my summons. There is no other feeling like it in the world.

THIRTY-EIGHT

WHEN SEX IS PARAMOUNT

The day after seeing my very first gray fox at Fox Hollow all those years ago, I went down into that part of the old dirt road and waited for a while, but saw no fox. I left feeling as though my sighting the day before had been a fluke and that I'd never see one again. I shrugged, but hope stirred in the background, patient.

My legs took me directly back to that old, galvanized pipe gate that crossed the road. Again and again, I trespassed. I learned how to stay out of sight. One morning, I stood there for no more than thirty seconds when I saw movement ahead, just a flicker. Something was there at the side of the road in the grass. I stood stock-still, waiting for another movement, when to my astonishment a fox walked out from the grass and onto the road. It sat there, like magic. A wild animal had come out to meet one of "us."

My camera chattered. I first saw that fox through my lens finder, because when the grass had moved, I'd snapped into what I'd thought was "expect anything mode." I took a few pictures and realized I was missing out on actually seeing the very thing I'd searched to find. The camera lowered and I looked. The fox, that perfect fox, wandered around, sniffing the air, the dirt, and probably picking up my scent.

Gray foxes are, on the whole, not much bigger than a large house cat or a terrier. For a moment, imagine yourself the size of a small canine like that, looking up from mere inches off the ground. Imagine yourself surrounded by trees, grass, blackberry vines, and ivy, and you need to know how to propel yourself through that blinding brush if an enemy or an alpha male decides to chase your tail. How do you know how to get from point A to point H to point Q and beyond, to your destination nearly a mile distant, and along the narrow, twisting paths beneath the brush? How can you possibly do that?

We human beings form a mind-map that begins with our awareness of being born: being squeezed down that dark tunnel, and an instant later, being shocked by light and sound and cold. From there, our mind-maps expand far beyond. As we grow, our mind-maps do, too. For instance, I know how to transport myself from here to East Glacier Park, Montana; I take a flight from Oakland to Kalispell, stay overnight at The Downtowner in Whitefish, catch the train called the Empire Builder, and ride to meet Bob up there on top of the Rocky Mountains. But I've done it for years, and I have all manner of technologies to point the way there and the route back home.

Foxes don't have the time necessary to practice escape routes or memorize ever-changing terrain, except in their early stages of life when they play and chase each other, all the while practicing to kill and develop their strength. Does their olfactory system guide them along their way, even when they stray from familiar scents to explore? Can they hear changes in the density of the brush? Is that how they form their mind-maps? What is their guide when they are in territory miles from their home range for the first time? How do they navigate through and into their unknown? They don't rely on a GPS or road signs, yet they seem to make their way flawlessly through any area, without a hint of hesitation, confusion, or backtracking.

Still, there are mysteries, many mysteries unfolding the closer I get. I wonder how foxes communicate across distances without making a sound—or do they? The only time that one can hear these foxes verbalize is when they are vibrantly happy or yelping in pain. Once in a great while, under certain circumstances, an adult might growl at a pup. They sing/call during mating season, during that time of year when sex is paramount, for it is a primal ritual that issues forth their litters of young, continuing their species. They call for each other, male and female alike.

Around here in the middle of February, it is that time of year when the chill of morning casts frost onto the weeds and grass. In the afternoon, the five pups, three female and two male, romped up on the truck washing pad. Pale was there, and Mama Bold, too. Gray lay off to the side, watching. Pale went over to the water and, as she drank, Gray came to her and sniffed around his daughter's tail. She went down on her belly before him, crouched low on the concrete, submissive, and trembled until he moved off toward Mama Bold.

Gray had done what most males do during that season, and the females all mimic Pale's submission. It's in their DNA. She comes into estrus or heat, her partner impregnates her once a year, and then she feels her little ones kicking her sides to get out.

I see Mama Bold's sides convulsing. Her pups want out, but before that can happen, she runs until she grows too tired. She stops, she sits, she looks around in the night, and she gives two rough, growling huffs followed by a low rolling growl as she sinks to the ground, moaning. I know this because my cameras captured video of the entire thing: Mama Bold is in labor.

Her body prepares to perform magic. She waits there amidst rivers of feeling that have come to her through time immemorial. In the next frame, she dashes to the interior of their natal den, that place she chose so long ago, where she sensed somehow that she would one day give birth to wet balls of fur.

Why there? Why not elsewhere? Why prepare? What kind of consciousness does that require? How did she and Gray know and agree that the alkali saltbush thicket

235

would be their home, instead of their original natal den?

As her body convulses, she feels a tight constriction that seems to go on forever, until at last, the pressure reaches its climax and she pushes her blind and deaf little ones out into the world. She looks down. She licks a wet ball of fur. As they escape her body, instantly, she knows each one's unique odor and each one's unique voice. There are five pups for her to lick dry, so tiny that she can almost lick three at the same time. Mama Bold turns one over with her nose and seems to favor it by licking with gentle strokes, slathering her daughter with affection.

That pup will become Daring. She is already being groomed for the task ahead.

Gray lies nearby and watches. With each new pup, his belly twitches. Later that morning, after they have taken their first nursing time, Gray goes to where his squirming little charcoal balls of fur slumber, blind and deaf, lacking for a time the two most important senses they will develop, the two senses alone that will allow them to survive.

Gray foxes can have a litter of one to seven pups, or so the research literature tells us. For the rest of the year, until about nine months have passed and the pups become adults, until they learn what it takes to survive and what it means to be a gray fox, during all of that time, the two parents will slave to feed and teach those growing balls of fur how to become competent gray foxes that will survive and reproduce in their own time.

It is no easy task.

THIRTY-NINE

LESSONS

Some say that they are solitary creatures, but they are not. They are instead individualists, but at certain times of the year, I see their need to be social and transfer information to each other. The males and females feed their young and teach them how to get food, but there's far more than that going on. The pups seem to learn the essentials of being a gray fox by watching the adults, and by way of instinctive play that is monitored and refined by their parents.

I stood one morning as Gray took all five of his sons and daughters into the trees for high-speed climbing school. I stood there and filmed it all, simply stunned by the power of their display.

It happened like this: I came upon Eucalyptus Hill late that morning. I looked for the foxes right near the natal den. The tall grass lay to my right when, as my heart quickened, the tops of some of that grass shook and Gray burst out into the open.

Five pups ran into the area, and Gray raced up a trunk to run along the branches high above me. Without a pause, the pups climbed up the eucalyptus trees in a mad, dashing frenzy of gray fox speed and agility. Gray led his pups. One by one, they went out onto the tip of a branch that barely held them, turned, and walked back along the branch with confidence and ease. Gray watched with approval and perhaps a touch of pride.

It only takes once for a fox to learn. They do not forget. The lesson in high-speed tree climbing was finished.

I have gleaned some ideas and opinions about the gray fox's behavior, some parts of which you will recognize if you own a dog and have paid attention. At other times this bush dog acts nothing like a dog but much more like a cat. It sits like a cat. It teases its prey like a cat. It climbs trees far faster than any cat, as fast as a squirrel fleeing

239

for its life from a predator. It is the only canine in all of the western hemisphere, down into Mexico, through parts of Central America, and into northern Venezuela and Columbia that can perform such a feat. When I first came to know these foxes, I called them "canines that act like felines" ("canine" should actually be "canid," but then it would lose the nice rhyme). The gray fox needs its own classification in the scientific rubric because it is neither a canine nor a feline, but rather its own unique family.

In many cases, the behaviors I observe are not found in research journals or in academic circles because no one has been with the foxes daily and watched what they do, as I have. I've called to them, even ordered them, "Come on, come on," and seen them pick up their pace to a trot until they come alongside me. I've watched them come from the brush on the levee road, in the very early morning before daylight, and have seen their silver eyes aglow down the road. I've called with a rolling timbre in my voice, "Good morning, good morning, good morning," and had a gray fox trot right on up to check in with the other foxes as it joins the group. I've spoken to them, or with them, "How are you this morning? Did you get a bellyful last night? A nice fat woodrat, maybe? What'd you eat? You still hungry?" I know that, like your dog, they come to know the meaning behind my tone of voice. They come to understand. They read the flavor of my tone just as much as I read their actions and voices. We have shared trust.

I know when Little One is going to pick a fight with an invading pup. I call, "Hey, Little One." She turns to me. I can add, "C'mere, c'mere" for a speedier response. She walks up to me. She understands my intent. Why don't I understand equally?

Their behaviors slice through what it means to be alive and to be a mammal. I wonder, do you remember that such a word applies to you and me? Are you aware that you are, at this very instant, a mammal? Does it occur to you that despite your clothing, and your smartphone, and your driver's license, you fall within the same classification as a gray fox, a mammal? Isn't that something? It may seem trite or unimportant to some, but I maintain that remembering such facts will keep us humble, keep us open to the possibilities of the wild.

FORTY

THE RINGER

Five foxes in the big clearing just relaxing. I sat on the huge rotten log there in that gap surrounded by tall trees. I scanned the scene: over there lay Dark Eyes, the alpha female of the region, the one I would come to call my Buddha dog, and off to my left lay the two sisters, Tense and Tippy. They hung together. Just behind them, up the slope, sat Big Guy with his legs spread, his penis showing. In front of me lay the female fox who'd given birth to him: Cute. When she had nothing on her mind, like that afternoon, she'd just flop down flat on the ground from stem to stern and close her eyes. I sat there listening, watching as the foxes occasionally stood up and moved over a way, then lay down again. I waited for them to separate, to choose the moment when this experience would end for me.

Instead, my cell phone announced itself. It was rather soft at first, but the sound rose in a crescendo with instruments, sounds, and noise filling the space. Every gray fox in that clearing was instantly on its feet. Dark Eyes focused, glanced over at me for a moment, then stared hard at the Ivy Trail, demanding to know what that sound was and why it had pummeled their solitude. It had startled all five foxes. They pointed in the direction they thought they'd heard the chimes and whistles come from, but they seemed disoriented. Not one of them keyed in on me or my cell phone as the source.

Now, that tells me something. They had never in their lives heard anything like that, and they had an innate need to find the answer to an unspoken question. They wanted to know about these novel sounds, to know if they held danger or pleasure. Does this mean that they have an understanding of epistemology, the study of what it means to know? The classic epistemological question is: "How do you know that you know?" Do they have a desire to know given their own paradigm, one considerably different from our own? Everything I observed told me that, yes, they do.

My cell phone was set to a musical alarm. When all that so-called music and clamor stopped, the foxes still clung to that stunned pose for maybe three seconds or so, and then they relaxed. Most went back to where they had been and lay down. Tippy and Tense moved closer to where I sat on that huge rotten log. I acknowledged each of them in turn.

Tippy came up close and put her paws on my thigh. Maybe she'd figured out the source of the sound, or maybe she sought comfort from me in the face of the unknown, or maybe she expected an apology for the disturbance. I don't know. What does her touch tell me about these gray foxes? What is the meaning behind experiencing such a novel event? What did it do for them, to them, with them? Maybe nothing. What it has shown me: I have yet to learn the lesson, but when Tippy sank her claws into my leg, there was certainly a physical connection. It felt magical.

FORTY-ONE

BURNING KARMA

It happened like this. My brother, Tom, had introduced me to digital photography, and I'd fallen in love with taking photos of all kinds of birds, especially the hawks—the predators. I'd even learned their names. The orange Bullock's orioles drew me like iron filings to a magnet. They were amazing birds, and a gathering of them had come in where those big blue gum trees stood as I walked down the old dirt road, on the dirt piece of Landings Road where the public was banned, down into what I'd later call Fox Hollow; off the road to my left, behind that alkali saltbush, was the gray fox's ancient natal den, though I had no clue it was there at the time.

In 2009, my gray fox addiction wrapped itself around me like a python, but this time for other reasons. I slipped past the "No Trespassing" signs that morning, those signs that actually meant nothing, but that certainly threatened an ignorant guy like me. I passed them, hoping I wouldn't get caught, and headed on down the dirt road to get some photos of Bullock's orioles. That's all I was doing that early morning, except that I was trespassing on city property, which could cost me—so the signs declared. I took a chance. I often do.

Brush on my right, I carefully walked down the road's edge and turned the corner, and there ahead sat a gray fox. As my brother, Lee, has said about my obsession with these foxes, "You're paying your dues, Bill." He sometimes says that with a knowing chuckle or smile, then continues with, "You're paying your dues, brother, burning your karma, for what you did to them foxes back when we were kids."

There was one particular place where those orioles seemed to hang out in the eucalyptus trees, but it was scary heading back there—not for the usual reasons you might expect, but because of the existence and insistence of *authority*. Video surveillance, invisible eyes, fines that spoke of penalties for violating city ordinances, signs three deep in some places.

NOTICE: This Area is Under
24-Hour Video Surveillance

Security Cameras in Use

NO TRESPASSING

In Accordance with Municipal Code 9.42.060

Trespassers Will Be Fined and/or Jailed

Suffice to say, the place was off limits for some man-made reason, so when I stepped across that boundary and trespassed, my old heart knocked and I was alert and finely tuned. If a truck came to the wash pad up the hill, I'd shrink away into the trees, out of sight. I made sure that no one saw me, but I could see them. That's the way of the fox, but I didn't know it at that time.

That morning, I came down the dirt road from the landfill and the park. I snuck through the coyote bush, weeds, and shrubs as I crept past the landfill's office buildings. I shook. I thought they were the enforcers of the no-trespassing edicts that blared out warnings from those black-and-white enameled signs. The truth, however, was that the signs were ultimately meaningless and ignorable.

I pressed on because there were orioles somewhere down there. Curiosity had taken me by my throat like a fox killing a mallard and drug me down that slope. I turned and my eyes set themselves on a fox sitting at the roadside, just beyond the galvanized steel gate of the landfill.

I stopped and drew in a slow breath. I hadn't seen one for more than fifty years. I didn't even know what kind of a fox it was just then. All I knew for sure was that I'd seen a fox like that eons before down on the Corralitos Creek, up by Old Man Dehl's place.

It sat there in what I've come to call the "regal" pose. It's passed down through the generations and looks like this: front paws perfectly side by side, front legs perfectly aligned and straight, chest full and out front, eyes partially closed in slits looking straight ahead, ears tuned in and pointed ahead, nose slightly raised. It's a striking image, at least to me. Some foxes really know how to present themselves.

This was the ancient gray fox, the fox whose lineage runs back millions of years. As some argue, that's far longer than we modern Homo sapiens have been on this Earth—somewhere around a mere three hundred thousand years, depending upon how you want to date it. In any case, we've just begun forging our history, whereas the gray fox has millennia on us. We are not yet mature in the pantheon of mammalians; we're still growing up and have so much to unlearn, so many mistakes to rectify. Some contend that we as a species won't grow any older, that this is as far as we'll make it.

The gray fox, *Urocyon cinereoargenteus*, has passed on traditions for far longer than humans have been in their present form on planet Earth. They've had plenty of time to gain every facet of what we have come to call intelligence, and probably more. What are their memes? How are they passed from generation to generation, and what geometry do they take? If we really want to understand these little bush dogs, and possibly find the key to our own continued existence, we need time—time to grasp the paradigm of their communication, so radically different from our own that we must start back at *tabula rasa*.

FORTY-TWO

MAMA BOLD HAD PUPS TODAY

I waited through the night and well into the following day, predicting when she would birth their litter, but as usual, nature moved at its own pace.

After calling several times and waiting, I saw Gray come back along the trail to their natal den. As I always do when a single fox appears, I asked Gray where his companions were.

"Where's Mama? Where's Mama Bold? Where is she?" Whether he understood my words or not, he knew exactly where she was; he had just left her. I called for her again. We waited for a moment, then back along the trail, the grass moved aside. Mama came through.

I could see her just about ready to emerge from the tall grass when she suddenly turned and went back out of sight. I'd never seen her retreat like that before; she'd always appeared when she'd intended to emerge, never hesitating or backtracking. I knew then that she'd given birth, and had probably gone back to nurse her young.

Gray lay in a small dimple of grass and groomed himself. I called a couple of times. Silence. Birds in a nearby tree flitted through branches.

Again, I caught the grass moving as Mama Bold returned. She broke through into the clearing and looked at me. I wondered if that had been the first time she'd ever left her newborn pups, and that was why she'd gone back: to check on them. Once she felt that they were safe, she eased away from them, found Gray, and nuzzled her mate, squeaking for him as he got up and walked away. Typical male gray fox behavior. She walked over to the edge of the slope and looked back at me as though asking, "You gonna follow me, or not?"

It was just then that I saw her underbelly, and it was obvious that her pups had been suckling. The fur around her nipples, those that I could see at least, was wet from her pups' saliva. Her breasts were enlarged with that gray fox milk her pups loved so very much, but which she gave only when necessary.

Her blind pups nursed for nearly ten days, experiencing a continual change as the world around them bloomed with changes in fragrance and sound that formed like waves. The pups had fun pointing their noses and ears in every direction, soaking in their surroundings as their mind-maps already began to form.

Their eyesight gradually accepted the light, dimly at first, but brighter and brighter until the pups made out images as seen through drifting fog. They saw their home, their mother, and each other. It must have been like an unfolding, a gradual becoming. Voices followed, and the den became a noisy place, filled with rumbling shockwaves from huge trucks passing on the road nearby, or tractors shoving tons of dirt away from the mountain to build a park for the people and the ground squirrels. The pups quickly understood that they could read the sounds, the smells, and the images all combined, and they settled in comfortably with the urban grumble out on the road.

FORTY-THREE

DAUGHTER PALE

When Mama Bold came to visit Pale, I'd just come over from the Fox Hollow area to Harbor Hill. I came down the slope toward a cluster of trees where my trail camera was positioned beneath the canopy; I needed to set it up for the night run.

I looked up and spotted Mama Bold through the bushes. She surprised me because she hadn't been there in months, but there she was, sitting in the clearing. I had just popped the trail camera open when Pale and Big Guy pushed through the weeds and emerged into the clearing. When Mama Bold saw Pale, she swished her tail, happy to be with her daughter once again, and Pale went into submission, coming in below her mother's chin for a fox-kiss. But then, without completing the gesture, Pale suddenly stood upright and met her mother as an equal. Mama Bold rolled onto her back and Pale sniffed her mother's vagina, possibly checking to see if she'd had pups.

Not yet. Pale walked to the other side beneath the canopy and lay there with an air of defiance about her. It confused me, so I took a picture of her. I had to ask myself, "What just happened here? What does that mean?" Was Pale jealous because her mother was going to have babies but she wasn't? She might have been pregnant if Blackie had still been around, but her father had run him off because he'd been an excess male. Pale had had to let him go.

As I consider it now, it feels like an old soap opera. Am I projecting my own drama into the situation, or is what I perceive the reality? Do these foxes ever live through the same scenarios as we humans? Could they?

I had told others with some great authority in my voice that gray foxes were solitary. That's what the research papers asserted, along with the peer reviews, but every one of them felt incomplete or irrelevant to what I'd observed, and my words never rang quite true. I was wrong. I'm okay with being wrong.

Gray foxes are independent, yet they have strong

social ties to some foxes, such as their mates. They are so independent that when they come upon one another, they have to re-recognize each other. They are sometimes startled by the appearance of another fox, one that they know rather well, even a mate. They sometimes flee into the brush or hunch their back, arch their tail, strut about with legs extended looking ferocious, but once they smell the other and know who stands nearby, all melts into a fox-kiss and they are at ease once again. They are independent, yet flexible. Each time they meet, it's a new day, a new gesture. Renewal.

Mama Bold came to disown Pale and leave her to her own ways. Gray had known that Blackie, Pale's would-be mate, had been a threat to their genetic pool. Gray hadn't allowed him to be in the vicinity when Mama Bold was in heat. Despite all this, Pale wanted to remain with her mother and father, but that was not the way of the gray fox. Mama had joined Gray in his attack on Blackie and helped force him from the area. The alpha had spoken.

FORTY-FOUR

CALLING

There were five pups that April in 2016. I scurried around trying to keep tabs on these balls of steel gray fur. It stunned me from the beginning that the pups were not at all skittish. Shouldn't their normal response have been fight or flight? Maybe those aren't the only options after all.

One afternoon, Mama Bold lay beneath the Italian buckthorn bush and watched me. It was a routine afternoon with my trail cameras and the foxes. I took several pictures of Gray lying there, then flicked on the trail camera and closed its box. The next morning, I'd find out what had passed that way, how many of each, and what they'd been doing within those thirty-second windows of digital files, all in high-def resolution.

From it all I hoped to gain a crystal clear understanding of these animals, but the truth is that I might only glean a thought or two that I hadn't had before—nibbles rather than the feast I craved. Even after all this time watching them, my knowledge is thin, shrouded, and unwhole.

Gray stood, yawned, and stretched just as any gray fox would do. He walked back into the thicket of their natal den there beneath the alkali saltbush. Maybe two minutes later, he came from the thicket and nudged Mama Bold, put his nose right into her side, twice. Slowly, the meaning unfolded. Essentially, Gray had said, "I just checked; our kids need to nurse. Time to nurse, Mama." Those nudges sent a message of some urgency.

He went back into the thicket. Bold stretched and gave her back a good workout, dipping low, then looked to the trail leading to her pups and disappeared into that tangle of vines and fallen tree branches. In my mind, I saw her standing back there, where some nights before she had pushed those pups into a totally new universe of experience, her legs slightly spread with five hungry pups, teeth sharp as needles, each suckling one of her six breasts. She stood stolid, erect, accepting.

I waited and watched. I saw Mama Bold's baby, a small bundle of fur, stumbling over twigs. Mama Bold came into the clearing and stood out there in the open. That little furball stumbled forward, half-hidden behind Bold, and looked up at me. It learned at that moment: "If Mama's not afraid of this thing, then there's no reason that it should frighten me—but it's so huge."

One by one, they learn lessons to gauge safety by their own degrees of comfort. Some were more skittish around me than others, but some seemed to trust me as much as their parents did. Each one is as different as a snowflake.

If I bring anyone who the foxes have never before experienced into the picture, then changes in behavior often arise. The foxes tend to keep a marked distance. They are hyperalert, skittish, ears scanning, noses slightly elevated, understanding the moment instantly. They watch from afar while at the same time sniffing and registering my friend's scent. *He is a new signature. He is a part of that one.* For the gray fox, that is how identity is established. I'm speculating, but I'd bet that my friend could return three years down the line and those very foxes would certainly remember him. It's not a stretch to say that their memories are absolute, eidetic.

Helper, the original helper female, always came up

to me unafraid. Every time we met, she stood there as I leaned down and extended my fingers, then she unhesitatingly approached my hand, licked my fingers, and nuzzled my palm with her wet, black nose. That act was her way of saying "Good morning." It was the particular way that we met, that we greeted one another. No other fox did anything quite like that.

Most of the foxes remained distant.

At the other extreme came the old number two helper, the second to arrive in the region. She was skittish and no matter what I might say, no matter what gestures I employed, she would not allow me to come close and stayed well away. I never gave her a name because one never surfaced for me, but I knew her by her tattered left ear. She had been around for a very long while, and had apparently fought hard. She kept her distance. It was she who would announce the death of the elders in Fox Hollow.

The pups grew up there in that small clearing where my trail camera was ready to record. But something went terribly wrong on June 18, 2016. According to my log, that was the last time I saw Mama Bold. Just before she vanished, I saw she had one breast that was engorged and black. It had been that way for some time. I doubted that the pups suckled that breast. Likely they could not draw her milk from it. What had caused the discoloration?

As foxes do at one point or another, Mama Bold no longer came when called. She vanished, never more to be seen, only remembered. That's typically the only indication I ever got that one of them had left the region or had died. They'd just stop responding to my voice, register as absent for a time, and then I would never see them again.

It's why I call their names so often.

FORTY-FIVE

FROM BAD TO WORSE

The death toll rose as more and more gray fox bodies were reported. The worst part of it was that when they first began to die, I had no idea why. I was helpless to stop it.

Cute wandered around at the corner of Forester and Quimby, not far from the creek. She staggered. The headlights blinded her. Stunned by the blow, Cute slid to the curb, twitched once, then lay still. I didn't see it, but I saw the outcome. It happens over and over again in my mind, even now.

Throughout December, I called and called, still unsure that she was dead. Hoping beyond hope that she was still out there somewhere, I called, "Hey, little fox? Hey, Cute. Hey, Dark. Hey, Mama Bold. Where are you?" I called and called again, "Hey, little fox. Hey, Tense. Tense, where are you?" and "Blue, hey, little fox. Blue, where are you?"

My voice crawled through Fox Hollow, rang down there in the overflow channel in the back of the city's maintenance yard: the virtual heart of Los Palos. For so long, I refused to accept that all those foxes I'd grown up with were gone.

That single event, Cute's death, sent a message throughout the gray fox community. It read that there was something evil happening within the family of foxes. Day after day I searched for her, but she never came. Gradually, my understanding began to settle in that all living creatures die, that all things pass in time. I released my grip on the illusion of control, and slowly came to feel that all was well, all happened as nature intended. I let them go because that was the way; the foxes had already shown me. I let go and freed myself to move on with my own journey.

Until I found Mama Bold's carcass: mouldering skin, exposed bones, her body covered with grass that lay over her like a blanket. It looked like she had crawled under that deep grass to die. The stench of the mother shook me. This was her—had been her—and now she was returning to the land, releasing herself back into nature.

Her mate, Gray, was not only a master hunter but was an attentive father, too. After Mama disappeared, he worked hard hunting for his five pups. I'd spot a black squirrel dangling from his mouth, woodrats, sometimes a bird or two, a duck (usually a mallard), a gopher snake, and, on rare times when the food lay scarce upon the land, Gray would bring home a Canada goose. He had become a single parent with the job of tending to his kids. As the scent of his mate faded ever so slightly from their home, I knew he would not let them down.

The second helper female had hung out in the area even before the pups were born. I'd rarely seen her back then. She'd been like a ghost, but once the pups were born, she'd come forward and helped Gray and Mama Bold raise their five pups. She was an old fox, her ears ragged from the many fights she'd had as she'd crossed through new gray fox territories on her path to Fox Hollow. She passed on the knowledge of the old ones to the male, Charcoal, and to the female, Slim: the last two pups out of the five.

FORTY-SIX

THE SIGNAL

It wasn't long before the pups came to the helper female, their proxy mother, their tails happily swishing as she approached, bellies rubbing the ground. Slim came up under the helper's chin simply in gratitude and to honor her elder. That's the way of the gray fox; all the pups approached her like that. When it came to me, she would not come closer than around fifty feet off, unless she was hiding under the bushes by the truck washing pad, in which case I could squeeze another ten or so feet in before she disappeared.

One warm afternoon, up there on the hill beside Sambeek's Shed, Slim sat over beneath the edge of the bushes, watching. Slim's brother, Charcoal, came up to me. I could tell by the way he approached, by the look on his face, that he wanted me to extend my hand, so I did. He sniffed my fingers and walked over to lay in the new green grass maybe three feet away, easy, relaxed, looking out across the concrete truck washing pad.

I hadn't noticed, but the helper female must have come from across the road, from her napping place under the coyote bushes, to be near the pups. I first saw her as she sat and looked directly in our direction, both ears erect.

Charcoal bolted upright, then stood, intently watching her.

In response, that helper female lay her right ear, the good one, flat over her right eye. Only that ragged ear stood upright. I had never before seen anything like that within the gray fox community. I had the impression that she was signaling something, that she was sending Charcoal a message of some sort.

Charcoal stared at her, clearly frightened and frozen in place. A beat passed and he finally turned and ran toward the back trail to slip under the brush. From that time on, the pup that had asked to sniff my hand only moments before would not come out into the open near me. He was skittish.

What had that flopped-over ear announced? Can

I take a guess? She'd announced that Gray and Mama Bold were dead. She'd told Slim and Charcoal that they were the only ones left of their family, and that they should be extra careful. I wonder, did she tell them I was to blame, or others of my kind? Did she instruct them to reignite their fear of humans?

In my log from 2016, I wrote, "Gray has died. He's been missing since November 5th, and the last time I saw him, he was blind." November 5th was the date when that helper appeared and sent her strange message.

Two days passed, and the pups were on edge the entire time. I couldn't find them or the helper female the next day, or the next. I never saw any of them again.

FORTY-SEVEN

GRAY: THE LAST TIME

The last time I saw Gray was the afternoon of November 5th. He came from the coyote bushes across the road from the truck washing pad, back in there where the helper stayed. He crossed the road, but he was unsteady. He sniffed his way along. I approached and looked closer. Both eyes were crusted over, each with a cake of dried pus. Gray was blind and sniffing, sniffing up onto the truck washing pad, sniffing over where the water puddle usually lay. I knew he was thirsty; the day was a scorcher.

Gray had suffered so many eye infections in his last two years that it had become commonplace to see him in that state of suffering. I'd tried to get him help, but

I've already told you about that. He was doomed. I felt sad that I could not help him, could not wipe his eyes so that he could see once again. I don't know what it did to his hearing or his sense of smell. The old hunting master of the region with his mate, Mama Bold, had ruled Fox Hollow from that day when Bold had battled with, and defeated, her father Squat.

I hurried over to the water hoses, caught one up in my hands, and turned on the non-potable water faucet. The water ran to the drain like a stream. Gray heard and probably smelled the water. He sniffed his way over and took a long, long drink. I talked to him: "Gray. Gray, I want to help your eyes. What does that feel like? I want to get rid of that mess coming from your eyes, but the people who might be able to help aren't interested."

I rambled on and on as he drank. It was my apology to him for the lack of humanity in my fellow humans, for eons of neglect and misuse of nature, for our failure to fill the role we'd assigned ourselves: stewards of the Earth. Those with the necessary resources really didn't care what happened to an old gray fox like Gray. If you'd asked them, they would have said that they loved wildlife, but that was all puffery; it was garbage they'd learned to parrot when they were asked to *do something*. Their actions never supported their words.

They had never been close enough to a wild animal to truly establish a bond, because such things don't lead to dollar signs.

That afternoon in November, after he had lapped up his fill of the cool water, he gazed up at me with an unreadable look on his face. I said, "Gray, I know." We were both helpless. He turned away and didn't look back as he slowly sniffed his way across the truck washing pad, across the old dirt road and off into the brush. I watched him go, and that was the last time I laid eyes on that great gray fox, the master hunter and loving father, that I had named Gray.

FORTY-EIGHT

THE LEGEND OF CUTE

About a mile and a half away from the truck washing pad, adult female, Cute, was the dumpster queen. I saw her dash up the nine-foot-high chain-link fence topped with three strands of barbed wire, reach the top, dive between those sharpened barbs, then bound down the last few feet to the concrete; just like that, she was on the other side. Her movements were absolutely perfect. Had she done it masterfully on the first try, or had she practiced for this moment to show me?

Cute trotted across the maintenance yard to a dumpster by the office buildings. It always had goodies, but there were other places to poke around, too. Often someone left an item, such as a pair of leather gloves, sitting out in the open, and when Cute came by, she would sniff, pick one up, and carry it away. She did the same with taco wrappers, a half-eaten roast beef sandwich in a plastic bag, and a carton of cold french fries; all would be gripped between her jaws as she trotted back over the fence.

The maintenance crew never thought about closing that dumpster because it was, at its top opening, some nine feet high. They all assumed that no raccoon, no fox, no animal living around there could possibly jump that high. They failed to recognize that the gray fox is, by nature, able to assess a situation, think through several alternatives, and act on the best possible choice.

Cute weighed her options and arrived at a simple conclusion: just climb the ladder built into the back of the dumpster. No one foresaw that the fox would simply climb that short ladder. That was something specifically made *for people*. No way some dumb animal could use it.

She was the only fox on the south side of Madera Creek who dared make that climb. She hauled baggies, burrito wraps, aluminum foil, leftovers from McDonald's or Jimbo's Diner—a half-dozen fast food brands—all back over the fence. You name it and it was there in one form or another. She littered the overflow channel with trash, with garbage strewn here and there along the way; it looked like a dumpster had disgorged itself. That's how it looked to me, but to the foxes it was simply delicious.

She would put as many things in her mouth as she could hold and head back over the fence. On rare occasions, her prize caught on the barbed wire and she had to let it go, for she simply moved too quickly to recover.

Her pups came running to see what Mom had brought them each time. All Cute wanted was to feed her pups in any way possible—except for nursing. She did everything possible to avoid those little snouts full of teeth. They hurt and made her underbelly bloody around her two nipples. She didn't have the requisite six nipples like the other females living there, and the pups wore hers out.

Cute's actions communicated a preference for juvenile life. She didn't like being an adult, yet there was no escaping it. Estrus was estrus, and it had far too many consequences for her taste. She could not avoid it. Anything that rang with commitment was blown to the wind. Having pups was a serious part of the game, and that's where she floundered and fell.

I had observed some time back that Cute couldn't tolerate her pups. In the three seasons before she'd had One Eye and Sideburns, she had but a single pup each year, and I saw that she avoided nursing as much as possible. Still, sometimes she had to give in. There didn't seem to be a reason for it, until I noticed the blood when she lay back and I could see under her belly. She sat there in pain from her pup's bite as he nursed. Her nipple was red, bleeding.

The following year, Cute and Dark had a single male pup under the Rose Den along the edge of the marsh, and I named him Big Guy. I set up a trail camera pointing across the brief clearing just outside the entrance to the Rose Den. The family passed by and became digital copies of themselves on my hard drives. In the video clips from my trail camera, Cute turned away when Big Guy tried to nurse at her nipple to get her warm milk. She left Big Guy, that little pup, standing there wondering why he had been so rejected by one he respected so very much. He went after her and tried to snuggle under her to nurse again, only to see her walk away in fear of his sharp teeth on her tender nipples.

Cute avoided being a supportive and teaching parent. Fear was etched into every cell in her body—fear of pain. The memes passed down through her cultural line must have been millions of years old, and I'm still uncertain whether her behaviors upheld or betrayed her ancestors' instructions. Whether she did as she was told or did as she wished, she ran from her parental role along with her aloof mate, Dark. That burning pain on her two good nipples wasn't worth conformity with the rules of being a good mama fox.

In terms of being parents as I think of them, Dark and Cute were the worst gray fox parents throughout the Baylands. In all my years of watching foxes, I've never seen a more negligent pair. That is my anthropocentric view of it all. I judge them, despite myself, but it could not have been easy for them. Were I to merge into their skin and peer out with their tools of perception, know what they know from their wealth of survival, and feel the pulse of nature within, perhaps I might have done the same.

FORTY-NINE

POETRY IN FOX LAND

These are independent urban foxes living in a quasi-communal setting. Normally they do not live like this, but human pressures and the destruction of their habitat force them to live unnaturally. There are too many foxes living in such a tiny space. Some foxes get along well with others, while some treat each other like they're invisible. However, there are those who come into a circle of foxes—

and bare their teeth
in a snap, shrill cries
slash, teeth cinched
precision calculated

shrieks, snarls
white belly
ears flat fury
protecting precious hearing
slithering across space
crying, "I have no animosity."
one gray fox slinks off
granting room
to breathe
to be

FIFTY

THEY DIED AS ONE

These were the gray foxes: the sisters with no names, and Gray and Bold, who lived a double tragedy of having their pups decimated, killed, murdered by a poisoned rat from over in the technology center next door, all because the people in charge of those buildings were seduced by the status quo of business, one that created suffering in the wildlife community of the Baylands by way of black boxes baited with anticoagulant poisons.

When I tell those who might make a difference about their poison, they give me reasons for why they will keep it. Even when these decision-makers opt to scratch their neurons, they seldom find compassion for anything beyond their bank accounts. Foxes don't hand out promotions, or land accounts, or push products, or raise the bottom line. How small and fragile, how *hollow* these words appear to my old eyes.

But even their poison couldn't match the effectiveness of the final blow that came in 2016, when the wind seemingly caught the canine distemper virus and blew it into every bush dog's body out there on the Los Palos Nature Refuge. That's the way it seemed: an epidemic that reminded me of reports about Ebola and how it had decimated so many African neighborhoods seemingly overnight. Canine distemper did the same to the gray fox community between the Hard Rock and Arroyo Creeks.

So quickly that it took my breath away, suddenly no foxes walked the trails beneath my cameras. None passed through Fox Hollow to play in the sun. None took the channel heading toward the marsh to get a plump rat. By the end of December, 100 percent of the gray foxes, young, old, and middle-aged, all perished.

They died as one, and the hills lay silent.

FIFTY-ONE

INCREMENTS

Dark Eyes was an enigma. When we first met, she "frightened" me in a way that I had never before known. She came from the brush with dark fur, almost black, with very little rusty color. Her gaze read me. I remember one of those first times when I found her intimidating, when that dark around her eyes seemed to glare like thick soot. She came from the thicket and stood no more than ten feet off, just staring at me, much the same as I stared at her. She held a presence that made me want to shrink away, to fall back, but I made myself watch and take pictures of her there in the big clearing. She was a wild fox driven by curiosity.

She showed up in the clearing more and more, but

she would not follow me nor any of the other foxes when they or I left to go out to the channel. She needed the canopy overhead, which is typical of a wild gray fox not fully accustomed to being in the presence of an urban landscape. She was not from around there.

We kept meeting up back there in the clearing. I kept to my work with my trail cameras as she kept watch, cautious of my every movement, but compelled to observe me. Over time, Dark Eyes relaxed. She gradually came with all of the other foxes out onto the overflow channel, following me, or stopping along the way if I called her name. It happened in increments, much more slowly than it did to the others.

I grew to expect those foxes to gather when I came along, and they did most of the time.

At first, I thought Dark Eyes was most likely barren. Even still, she commanded universal respect from all of the other foxes. That's not to say that being barren meant a loss of respect for her in their eyes; that's a human construct, certainly outdated in today's world, but foxes see differently than we do. Later, I realized that she was not barren but had pups when she wanted to, whereas Little One had pups each season like clockwork.

Dark Eyes commanded attention. She was the alpha female, and all of the foxes in the region knew her and showed their respect. Even when she met an elder male like Blue, he came to her, hugged the ground, slipped along with his belly sliding—well, you know the ritual by now. It's played out in so many situations by gray foxes living thousands of miles distant, yet it's always the same. It is innate—common in the wild but rare in captivity. It is a piece of the fabric of their culture.

Some will object to my use of the word "culture" when referring to a wild animal, presuming that such social constructs are unique to the human experience, but I couldn't disagree more. Here again we must alter our perceptions if we are to ever understand.

Down along Madera Creek, on the south side, lived an enclave of gray foxes that mixed, hunted, played, and mated. When two of the females were to have pups, have their litter there along the creek, for the first time ever they shared an area where two natal dens existed nearly side by side. As the young matured alongside their neighbors, their intermingled lives began to flow. The adults found that they didn't have to repeat hunting techniques, because to teach one was to teach them all, and in both families. The woods back there in the big clearing became one sprawling learning field, populated with raccoons, striped skunks, squirrels, and woodrats, all overseen by the region's keystone predator: the gray fox.

FIFTY-TWO

ON CHOOSING A MATE

It was well into the season when they paired up. The environment was open, accepting, ready to orient itself toward procreation. That's when a female gray fox selects her mate—makes it known that he's the one. When chosen, he is no longer free.

It was like that with Mama Bold and Gray. Bold was into her second year, right on time for a normal young female to select a mate. I hung out down there in Fox Hollow, watching and documenting everything.

It just so happened that, in 2012, there were many yearlings taking to the paths, moving through the Baylands, dispersing, and coming in through Fox Hollow. Like a slow drumbeat, I noticed that there were unrecognizable, unnamed, frightfully skittish foxes coming into the region. All of them found that Bold had claimed the territory; after all, she'd fought for it and won.

Sometimes young males hung around for less than a day, while others stayed for two, maybe three days, each checked out by Bold before they moved on, unwanted. Bold must have shown each one of those males that they didn't suit her, so they continued traveling on through the Baylands to find that special mate.

There came into the area a pale gray fox—by that I mean his fur was far more gray than tinged with rusty red. His nose seemed far too long, at least compared with the other foxes. His frame was slim, but with a broad chest.

He stayed around for maybe four days before I noticed he was no longer present, no longer following Bold. I took him to be a transient. She had traded him in for another young male that had come along, but that one was gone within a day. Bold must have gestured, must have somehow conveyed that pairing up with him just would not work. I wonder how she managed it, and what form the rejection took.

That choice had everything to do with genetics, those ancient pulses that define being alive through time immemorial. Add to that the culture of the gray fox community, the pressures of raising a litter of pups into adulthood, of carrying them through to their dispersal, of ensuring their care and feeding, and we begin to see that choosing a mate is not as simple as it may first seem.

When I first saw the gray fox family unfold, I had little idea as to what had just happened. I had just begun to bond with the wild.

Dark, the alpha male of this home range, lived without challenge. His parents selected him from their own litter as the one who would be alpha, the one who would quickly learn, the one who would be expected to perform certain duties as conditions dictated. That one emerged from the thicket out onto the road, and ventured forth out of authority tinged with curiosity. He was the one who stood there looking up at me.

He was the enforcer, the male that balanced the system's dynamics.

Nonetheless, Dark lived alone, aloof from all others, even his mate. Dark shivered amid the other gray foxes; unlike most that blended in with ease, he felt an emotional distance, brought on by his station. Cute came to him, showed him her vulnerability, gestured in every way she knew how that she wanted his attention. Mama Bold did the same. Most often, Dark walked off, leaving them there lying beneath the alkali saltbush. On the other hand, Gray stayed close, occasionally nudging Cute to remind her that he enjoyed being with her.

Dark reluctantly helped Cute raise their litter of one pup, but he had another duty to perform, a meme he had learned from an elder. This impulse grew from having wisdom passed down, not through the genes, but through their culture. Those stories displayed through gesture, through sound, through experience. That culture called to him, saying, "As the alpha male of the region, you must remove Blue, and any other intrusive male gray fox, from the home range. You decide when to act, but failing to do so will put us on a deadly path."

FIFTY-THREE

TRAPPING THE GRAY FOX

A fox lived in our prison, awaiting its sentence. Many others had come and gone, and there would be many more after this one. The fate my brother and I envisioned was to sell each fox we caught to a man that Dad knew, some guy who owned chinchillas and raised them for fur. Dad said the guy would come and buy our foxes, but he never showed.

Sometimes, probably in utter panic and a desperate need to be free, the fox chewed its leg to the bone to escape from the steel jaws of our traps. There was no such thing as padded or so-called "humane" traps back then, just metal, flesh, blood, and bone. The traps were numbered according to levels of strength. I no longer remember which kind we bought, but they definitely worked.

We had several foxes break their legs trying to escape, and my brother Al and I had to take care of the blood and bandage them so the wounds could heal...or so we hoped. Since then, I've seen a remarkable healing rate out in the wild without any human intervention. They have astounding resiliency. Still, the damage had been done.

Nevertheless, within a couple of days, injured or not, every single fox escaped, and with each escape I learned how much smarter they were than my brother and I combined.

We were just kids with a weird hobby, making things up as we went along. No one cared what we were doing because no one knew anything about those bush dogs. They were just "critters" to everyone.

Each day that a fox escaped, Al and I spent hours down there on the cage trying to make it fox-proof, until we simply ran out of ideas. Despite the best efforts of two boys, the foxes won every time.

Years later, I learned a little more about their intelligence. But what does that term mean, "intelligence," and how does it relate to the gray fox? Rudimentary intelligence involves the ability to know that two independent and disparate things exist at the same time, and that a change in pattern occurs when those two items are put together. Additionally, foxes would necessarily have to understand abstracts like duration and time, and they do. That is the essence of being a gray fox. They tell time in an organic manner; as winter moves toward spring and the length of sunlight extends, they have varying issues that they must deal with, like the shift from feeding on a rodent to bringing back plums to the den. Throughout the year, they are adaptable, able to merge with the flow, enjoying the ride with infinity.

But to make a statement like "Foxes are intelligent" suggests that I want to measure fox intelligence with the tools we apply to human intelligence. Maybe that's not fair. Maybe we blind ourselves to alternative possibilities. How else can we measure intelligence when we have no concrete idea as to what it is, except to say we recognize it when we see it? Why do we think we know anything about it?

When I walked together with these foxes, I saw intelligence far greater than our own. We cannot compare.

FIFTY-FOUR

GOPHER FLAGS

The new superintendent and some of his crew watched it happen. This is my approximation of what Stan, the head cheese, said: "When I took over here at the course, the greens out there were full of gophers. Most of 'em, anyway. The crew before us used poisons to try to control the gophers. Didn't work, ya know?" He shook his head. "When I got here, I knew they'd failed, and besides, it cost the city a hell of a lot for that poison. I told Joe, 'Let's go back to the good old trap.' He agreed, and we were set."

Stan told me, "Monday morning is orientation here at golf course maintenance, ya know? I'm the boss. When I announced traps that Monday morning, back maybe six months ago, all I got were looks of disbelief and real grunts and groans. I'd never heard that from my crew before. They were pissed. I saw it. I felt it right down here." He grasped his plump stomach.

"Yuh mean—uh, yuh mean we gotta get out there every morning and check all those *fuck-inn* traps? How many?" Ruben, the gopher go-fer, asked with his undeniable emphasis.

"Johnny thinks 'bout two hundred and twenty-five or so, give or take, ya know?"

"And we gonna check two hunnerd and twinny-five *fuck-inn* gopher holes every morning?" He sucked in and let slide, "I'm not. You?" He jabbed his thumb off to the side, off to his left, and went on, "I'd never get the rest of the groomin' on the greens done, and then we'd have those golfers breathin' down our necks, and then we's fired and got no work. 'Sides, you cain't make me do grunt work like that. City wants us off poison. Doesn't work and we all know it."

Nevertheless, for two months—two long, griping months—the crew checked all the traps (or so they reported) because Stan didn't know what else to do. The problem stewed and churned, and gradually, something jelled. It came to a head when he finally couldn't take it another morning and called a meeting in the lunchroom. After his orientation, he announced, "Hey, gophers, so I got an idea." The crew stood silent, skeptical. Stan pointed and said, "With every trap, we rig it special with a flag—a red flag. Got a lot of 'em over there, some promotional thing's leftovers. So when that gopher's caught, when the trap springs, the flag pops up. That way you don't have to go check every trap, only the ones with a red flag, and you'll see that from a long way off while you're doing your regular work."

Human ingenuity. I wonder how the foxes would have solved that problem.

As Stan commanded, so it was. Red flag up, a gopher in the trap, or at least that's what the crew got for the next four days.

Ruben burst in. "I's jus' checkin' that trap with a flag stickin' up. Only thing ah found was a bit o' fur and no gopher! Trap was outside the hole, too."

Stan waved him off, turned to me, and said, "Within the week those foxes learned that trick. They're savvy, huh? They knew where the food was out there 'cause the flag was up. They saved us having to dispose of them gopher guts, and the guys were happy about that, too. I tell ya, foxes ate them gophers."

I said, "Wait a minute, you mean to tell me that those foxes learned?"

"Yeah, they got it figured out. They learned that when a red flag flew across the golf course, there was food, and they trotted out in the night and picked up gophers from the traps where they were."

"But they can't see red. Must have just been that they saw the flag, and flags meant gophers."

"Either way, they sure made our job a lot easier."

FIFTY-FIVE

LUNCH CART

It was lunchtime in Silicon Valley and many mid-level high-tech managers still held to the notion that they might do business on the golf course. Critical get-to-know-you meetings came together out there on the greens. The Bayside Cafe and Restaurant would put together boxed lunches for the golfers on their noon round of nine holes, or if it was an extended meeting, all eighteen.

Most of the time, the foxes already knew which golf carts had food; they were easy to smell. This time the gray fox ambush took place next to the airport. That was the fifth hole. Just on the other side, a wall of tall shrubs and bushes skirted the green. The golfer, with his lunch there on the seat beside him, drove up.

Three foxes watched from beneath the bushes off to the side. The alpha female of the region knew the exact moment when to dash to the golf cart, snatch that lunch, sprint across the open space, and zip back to the safety of the fence.

"You stupid damn fox!" screamed Randy, cofounder of EgZact Technologies, his voice cracking slightly, furious that he'd just been robbed once again by *Urocyon cinereoargenteus townsendi urbanus*.

Instead of just standing there in shock, Randy dashed to his electric cart and aimed it toward the bushes where the fox had vanished. He punched that pedal to the floor and the cart bounced off the chain-link fence. He jumped out and scanned the edge of the thick bushes, kicking his foot through the foliage here and there. Nothing stirred. He looked into the brush, spread branches with a gloved hand, got down low on his knees, and peered back under the bush's canopy, but he saw no fox, no lunchbox, nothing worthy of his time. Exasperated, he huffed and strode back to his cart, backed it off the fence, drove it to the cafe on the green, and bought another sandwich. This time, however, he decided to eat in the clubhouse.

The urban foxes have discovered a new facet to

their survival: humans can be avoided, but they have all the best treats. They are a danger, but also an asset. Gray foxes have learned how to live side by side with humans, and such behavior goes deep into the past with the Native American tribes that lived in the southern reaches of Canada, along the Great Lakes, south across America, down into Mexico, and on a delicate strand into South America. Some tribes in California revered the gray fox as a god. *Ketchimanetowa*: the creator-god of the Fox Tribe. It is also known by other names: Keckamanetowa, Great Spirit, Atius-Tirawa, Gitchi Manitou, Great Manito, Great Manitou, Ilex, Kisha Manido, Kitanitowit, or Manitou.[2]

FIFTY-SIX

OUTFOXING THE FOX

Steve and I are in the same business. He wanted to show me the red foxes of the Los Lagos Golf Course in South San Jose. On our second time back, one of the trail cameras that I had loaned him had been stolen. No big deal, but Steve wanted to take me back out on the range and show me the foxes again, so of course I went.

People call me the Fox Guy. I have been studying the behavior of the gray foxes around the San Francisco Bay in California for twelve years.

At this time of year, the foxes, both red and gray, are denned up with their mates, ready to have the year's litter. If a human comes too close to the natal den, the male will intentionally lead the person away, remaining just out of reach but close enough for them to follow. Once far from the den, the fox heads back into the brush, slips past their quarry, and returns to his mate and their den.

In this case, the den was high up the hill above us.

Steve pointed to where dirt fell over the edge and down the steep slope, saying, "The den's up there. See it, where the dirt opens up?"

"Yeah." And right then, a red fox came from the brush, charging downhill from up near the den. It ran past us, maybe thirty feet off through the brush, and out onto the golf course road.

Steve barked, "Come on, come on, let's get some pictures! It's going out there on the golf course, I see him. We can get some if we hurry."

I cut him short. "Look, Steve, no. That fox is going out there on the road to lure us away from its den up there." I pointed through the trees behind us. "He thinks we'll follow. That's what most people will do." I paused a moment to gain my bearings, then spoke as I moved my finger around in an arc. "He's going on around and back over there to the creek, and he'll come along down in the bottom, cross over, and head back for his den. Come on, I'll show you." I motioned to Steve. "Let's get right over there."

I led the way. I hadn't taken much more than twenty strides when I topped the backbone of the ridge. Steve stood just behind me, and that fox ran toward us right up the trail. I hadn't expected it to be that close. When it saw me, it froze. Shock blew across that fox's face as it turned and ran down through the brush, across the gully, and up the slope to its den.

I turned to Steve with a grin on my face. Steve burst out, "Oh, my God, my God! The Fox Guy just outfoxed the fox!"

FIFTY-SEVEN

MUD

It may have been tucked off in the corners of the Baylands, waiting. Not enough wildlife died to call anyone's attention to it—just one or two raccoons, or a single gray fox in the middle of Landings Road. The foxes looked healthy. Oh, once in a while, they'd get injured when they made that arching leap into the tall grass to catch a rodent, but their immune systems took care of any damages. I was always amazed at how they healed from what appeared to be serious injuries.

I asked for help early on, but no one showed up. Not even the rangers.

One day, I noticed that Gray was pawing at his left eye. It looked infected. There was a little pus-like ooze in the corner of it—ocular discharge. I thought, "Maybe a foxtail got in there and became infected." I convinced myself right away that it was nothing, that it would heal in no time. After all, he had healed up pretty quickly from multiple bloody fights. All their millennia of survival had given them robust immune systems, so of course Gray would shake it off.

After nearly two seasons of the foxes scratching away patches of fur where vermin burrowed into their skin, the foxes' fur filled in and decked them out with their late summer, early autumn coats. Mama Bold and Gray's pups were no longer little fuzzballs and now held their true colors. The one I called "Slim" looked the most like Gray: lighter fur and no bridge on her nose, nearly flat all the way to the tip. Mama had splayed ears, and two of her pups inherited that trait. The others had fairly tight ears, closer together and much more erect, like Pale's. Some of the pups displayed a unique characteristic or a certain behavior that was different from the rest, like the one I called "Shy," who decided to separate herself from the others. She carved out her own space back along Fox Hollow Trail, while all of her siblings denned up by the big pipe that fed the saltwater channel. She seemed to have a strong independent streak.

But she broke when she put Daring in her place.

LOG PAGE 305—Monday, April 17, 2017:
This morning when I went over to the den area, the gray fox pup Daring looked ragged, her fur matted, just not right. She didn't come from the weeds so I couldn't get a good picture. In the video files of the den site, there is a file #Trail Camera—06-05 & 06 2014—#1 Den area EK000003—Muddy gray fox pup Daring & Bold trying to help. In that file, the pup Daring has mud everywhere, and Bold appears to be trying to help. For an instant, Bold looks like she wants to get the mud off Daring, but Daring runs. Bold runs after her.

I don't know what happened. My trail cameras couldn't record everything. The next morning's video showed that most of the mud seemed to have vanished from Daring. Had Bold licked it off, as she had when that pup had first been born? How had Daring been cleaned up? And what had caused her to get so muddy in the first place?

In my mind, Daring had most likely been trying to take down a mallard out on the edge of the saltwater channel, and when she'd rushed in, the mallard had shot into the air, causing her dive to miss the duck and meet the mud. She'd simply missed and paid the price.

FIFTY-EIGHT

MOVING PUPS

Three unusual seasons passed—spring, summer, and fall. The drought of 2016 ravaged the land. There was not enough fresh water in the vicinity for the gray foxes, but somehow water oozed from the hillside and drenched Fox Hollow. There was no cause and effect there, only coincidence. Water bubbled from the ground, flowed across the road, and created a freshwater marsh that sat alongside a vast salt marsh.

Frank sent a few of the junior laborers down there to find the leak and stop it. They dug, only to have their holes fill with water, and to have no answer for where all that water came from. They dug across the hillside, and water flowed from the dirt into each of their test holes as if it came from everywhere at once. The central hole filled with water in the midst of the drought and became the salvation for all manner of wildlife: birds, hornets, honeybees, frogs, and more. Soon there was a small freshwater wetland that had magically shot from the earth and now blossomed with sounds rising from every direction.

It was alive—a temporary paradise.

After three months, salvation turned to mud. Wildlife drifted elsewhere for water, but in those days, there was nothing save for the saltwater flowing into the channel. The street sweepers would come and wash down on the truck washing pad; once they rumbled past and all grew still again, then came the opossums, raccoons, and skunks to gather around the trail of shallow puddles for a sip.

I watched. My trail cameras watched. I begged for help—but you've heard all this before. We set up that large tub of water so that all of the wildlife could drink from it, and they did.

The drought of 2016 came down hard on all of us.

Mama gave birth to five robust pups down under the alkali saltbush. I don't know where she got her water, but maybe it came through the bodily fluids from the rodents she ate. Her pups nursed. She winced when one or two bit down too hard.

One morning, I sat at my computer, downloaded the images from the SD Cards that I'd gathered earlier in the dark of morning, and loaded them into a folder, excited to see the images that never ceased to surprise me. There in black and white stumbled a dark gray, fuzzy pup, tripping over branches and sniffing with his parents in tow. The whole family appeared on my trail camera at once.

I laughed with pleasure.

The pups grew, exchanging their charcoal fur for rusty red around the ears and down the legs as they took on their adult camouflage. As they did, Mama Bold had a tradition that she'd pass on: she'd move her den every fifteen to twenty days. Over the years, she tended to return to the same areas for dens three, four, or however many times she decided to that season. For instance, after the natal den with her first litter, she moved her pups over to a grassy area where there were twenty-foot-long polyethylene pipes, and that became their next den, their next playground where they would learn the art of becoming full-fledged gray foxes—right by the big pipe that brought saltwater from the San Francisco Bay into the Saltwater Wetlands.

Did Mama move her pups to avoid the age-old fear of a predator, driven by that maternal need to protect her pups? I don't know, but I do know that the following year she moved her pups down into the same area. She found the place slightly changed: the pipes had been moved, the grasses weren't clustered enough, and the cover wasn't just right. But nearby, a large, red shipping container had moved in just up the slope near the big pipe.

Mama Bold took her litter to live beneath that red container, so that's where I went to observe them. Three hefty little pups dashed under the container's steel edge, and they hid and slept under there. I waited and watched until a pup's nose carefully emerged, then half a pup edged out further, listening, hearing, sniffing, smelling all the information. Data in, thought out. Confident, the pup stood there looking at me, my Canon digital camera planted on my face as I took pictures, hoping that I could get enough shots before they moved again.

They played and lazed around that container until they were rather hefty, putting on their adult weight, gaining more and more adult color, and easing into their adult statures and statuses. They developed in their own ways. There were differences.

Five pups moved out, increasing their range night by night, dispersing into the countryside, driven by curiosity. Eventually, they moved up the road from Fox

Hollow, out to the coyote bushes and beyond, to places I couldn't follow.

FIFTY-NINE

NAMELESS

From here on this story becomes extraordinarily complex, so much so that it's hard for me to get my mind around it, much less find my way through the spirited soup of gray fox relationships.

I've mentioned a helper female hanging out with the litter, but sometimes she'd also take the pups away for maybe a week or more, and then return. I took it for granted that she was just expressing her role in the Fox Hollow gathering.

I don't know when it happened, but I lost track of the last time I'd seen the full litter. The helper took them out and brought them all back—all but one. Then she did it again, and again. That pattern played out until we were down to two pups: Slim and Charcoal. All the rest had vanished like the last time I saw Gray.

The last two were memorable: Slim, a young reticent female, and Charcoal, a young male who fearlessly walked up to me while his sister watched from the bushes. The last time I saw the two of them, Charcoal came and licked my hand. Slim wanted nothing to do with me. Her right eye was infected; it oozed that yellow pus-like drainage. Even then, I thought that she had just injured her eye. I had no idea that she was infected.

Before that sequence, however, Mama Bold continued nursing their litter. Their pups had been born in April and it was already June. Routine dominated. The pups came to know me and they weren't afraid.

The helper female appeared. She was so wild and skittish that she remained out on the periphery, and she remained that way until I no longer saw her.

I came back out along Fox Hollow Trail. Off to my right, something shot off through the tall weeds, and suddenly a gray fox scrambled up the chain-link fence. She paused an instant before sliding over the top and down the far side, then vanished. Later, I found that she had injured her paw, and it was likely what had caused her to pause when she'd gone over the top.

The pulse and the rhythm out there radically

changed, and not just at Mama Bold and Gray's home range. Over on Madera Creek, Dark looked frightened whenever he came from the brush. When Blue showed up, Dark knew that he needed to chase Blue as far away as possible, but in that last season, he lost all will to actually do so. It was like Dark gave up. He hung out near Madera Bridge, and whenever I came down and called, he came from the brush almost reluctantly, then skittishly fled to the opposite side of the channel. That was not his nature, and I could tell something was off with him. Cute only showed up a few more times before I never saw her again.

The ranger got a call about a gray fox being struck by a car over at the corner of Forester and Quimby, right in the middle of houses and streets and cars, but not far from the creek. Gray foxes were showing up dead everywhere in the Baylands. I got calls or emails from the ranger each time to let me know that they had another dead fox out in the cooler; they'd ask if I wanted to come take pictures before they disposed of it.

They were sand through my fingers.

Of course, I needed to identify these gray fox corpses, but I could not tell which fox was there before me, no matter how hard I tried. The nature of who they were, as foxes, as individuals, was intrinsically tied to how they'd moved and acted in life. In death, they were strangers. I realized that my relationship with them had become much more than simply knowing each particular animal. I could not identify any of those foxes that I had named, had come to know so well, and had known me as I'd known them. When asked, "Do you know this fox?" I had no answer. Such immobile, inexpressive forms held no resemblance to Gray, to Cute, to Blue. They were expressionless cadavers.

Whenever I was called in to identify one of the foxes, I always walked away uncertain. I'd look and look hard, trying to find one of those marks, one of those identifying physical features that I'd come to know so well, but in the end, I understood that they were nameless, and I wept at the loss.

It happened so quickly, but that's the way events manifest out there. I hadn't seen Mama Bold in a while, so I called.

Mama didn't show.

SIXTY

DEATH ROAMED

A Cooper's hawk kept trying to snatch one of the pups, even though the pups were the same size as it. One afternoon, I came down the old dirt road into Fox Hollow, and my eyes settled on that Cooper's hawk just standing on the ground to the right of the trail, the one that the pups used to go back and forth to their daytime warm spot in the weeds.

Our eyes locked.

The hawk froze for a split second, then flew away, but that wasn't the last time I saw it.

Now there were only four pups, and Mama no longer came around to nurse them. They hadn't seen her for days. They must have realized that something unsettling had happened. They shivered, but didn't know why. The pups soon learned that their mother lay out back there in the weeds, under the grass where she'd crawled next to the marsh, never to move again. One by one, they each visited her, and then they left. They moved everything over to the hill behind the truck washing pad.

I hadn't found her yet. I didn't know that Mama Bold had died, but her four pups knew.

SIXTY-ONE

IN PASSING

No warning. No perceived indications like being unsteady, but on occasion, I'd see one of them cough from way down deep in the belly.

Mama Bold was gone and three of her pups had vanished, leaving Gray with the final two. He went to heroic extremes to keep the last two well fed, while at the same time teaching them about being a gray fox. Mama wasn't there to add her polish. Most of his teaching came through doing. Most of their learning was to mimic.

Pups learn through example and then embellish upon it. I sometimes ask myself, "Do they ever forget anything?"

Gray was a remarkable male, and I clearly understand why Mama Bold chose him instead of the others that came through back in 2012. I want to think that Gray knew that his longtime mate, Bold, had died back there on the edge of the marsh. He took on both her job and his own. He let the kids groom him and he groomed them in turn. The helper female, the one with ragged ears, became a part of the family.

She sat over there on the truck washing pad. Her right ear flopped over, limp, most likely due to a fight. The pup I called Slim hung back under the bush, and her brother came out. He stared at the helper, sitting over on the concrete slab, and suddenly he shuddered and charged back beneath the brush for the last time.

SIXTY-TWO

WHAT WE FAILED TO SEE

For two years, Gray's eyes had drained with each infection. I was too ignorant to see what was happening. I had but fragments of a landscape that defined the greater reality. All engaged should have known that such drainage from the eyes is a symptom of canine distemper, but we didn't know. Even those who were schooled in such things didn't recognize it.

For those two years, Gray's immune system must have hammered at that virus. His eyes healed and he saw clearly for a month, maybe more, but then the drainage returned and cast him once more into the depths of lost eyesight. What a loss that would be for anyone. He suffered. The ebb and sway of Gray's immune system determined whether he was blind or whether he could see.

Gray's eyes were pasted over with a pus-like drainage that, when dry, sealed his eyelids shut. He was blind— such a contrast to his younger self. From the beginning, whenever I'd followed him as he'd hunted, he'd caught upward of 90.8 percent of the field mice he'd gone after. That percentage is stunning, even in the realm of the fox. He truly was a master hunter.

I hadn't seen Gray for awhile. He didn't come from wherever he was lying in the late afternoon, even when I called his name. It was hot that afternoon, up in the mid-90s. In such heat, the foxes lay under the willows, back in there where it was relatively cool.

We've been here before, but the memory calls me back.

Again I said, "Hey, Gray, little fox. Hey, little fox, you there?" Across the truck washing pad, from the road over by the coyote bushes, Gray came across and up to the concrete slab, sniffing all the way. I saw that both of his eyes were pasted shut and shook my head in dismay, realizing that he was blind once again—not only that, but he was thirsty. He sniffed toward where there was sometimes a puddle that he'd drink from, but it had long ago dried from the hot concrete.

I thought, *He's thirsty. What can I do?* Just behind me were four reels of hoses, ready to spew water for the drivers to clean off their trucks with. I grasped the lever and pulled one toward me. Water slowly managed to move out through the hose and down to the drain. Gray came to the sound of the water on the concrete and there he lapped at the spill. Once sated, he turned back toward the coyote bushes across the road.

I called to him one last time, "Hey, Gray. Where ya going? Hey, little fox." He should have turned to look back, as he usually did, but instead he sniffed ahead, moving away. I wanted Gray to stay for awhile, to be near me and let me know everything would be all right, but he walked to the thicket and disappeared. Hindsight confirms it now, but I knew then that not all was right with him, and he knew it too. He died back there in the brush, not far from his mate, Mama Bold.

It's Monday, December 26, 2016. I think I finally found her remains today. I'll never move them. I looked for Gray, for any of the missing young ones, even for the helper female who had worked to raise those doomed pups.

Along the way, back there in the brush, I slogged through the deep, dead grass. Beside the pickleweed, I smelled death. I smelled that rotting flesh and bone, present but just beyond sight, beyond the capacity of my limited senses. I looked and looked, but I never found the source. It may have been Gray or one of the pups. I don't know.

SIXTY-THREE

ONE EYE

But then, back there in time, months before, there were those two pups over on the Madera Creek overflow channel, the one I called One Eye and his brother, Sideburns. These two were oddities from the beginning. They were slow- to non-learning pups, and they had no idea what a gray fox needed to do to remain alive, except that sometimes they saw a hungry adult bound back into the woods and get a fat woodrat, then bring it out on the grass to chew and rip and savor the flesh, coated in blood.

These two pups may well have been the result of incest between Cute and her son, Big Guy. They had no idea how to survive. No adults had ever taught them. It looked to me as if Cute and Dark had simply abandoned them.

They lay in midday out on the dry grass along the channel, and I watched to see if they would wake, or move. Normal gray foxes don't just carelessly lie out in the open like that. Normal gray foxes lie beneath the canopy.

One Eye had had eye infections from the beginning. He became the key to unraveling the mystery of the pus-filled eyes. He hardly knew what it meant to see from both eyes. I didn't seek to find a cure for him, partly because I thought that the eye infection would heal of its own accord. I'd seen it happen with Gray, who lived about a mile and a half away.

However, I felt agitated knowing that this eye infection had shown up so far from Fox Hollow. I traveled through fox-land day after day, wondering how it came to be in the overflow channel. I had no answer.

Time went on, and I noticed that One Eye had a very hard time either defecating or urinating. It took him minutes, or sometimes much longer. I timed him on my stopwatch, just to see how those bodily functions worked. It once took him six minutes and thirty-three seconds to defecate a small scat. Imagine that for a small pup. The distemper virus had taken him, too, but I didn't know what was wrong at the time.

One Eye disappeared. The last time I'd seen him, he was very sick, hardly wanting to move, head drooping there by the ditch. I have a picture of him somewhere, but it's hard to look at. It felt as though the gaping jaws of nature had opened and slowly consumed him, until that email from Animal Services said that they'd found a gray fox pup hunched in their driveway when they unlocked the gate coming to work that morning, like he'd hand-delivered himself to them.

The supervisor sent me that picture of him before he was shipped off to Burlingame. It was One Eye, all right, and they took him in. He harbored a mite infestation, and that meant his immune system was under attack. He was sent to the animal hospital for wildlife to be healed, to recover, and to one day come back to live in the channel where he'd been born. That was the plan, they told me.

Over the following two days, he grew so sick that they had to euthanize him.

Unofficially, the word throughout the Humane Society was that his breath had smelled like distemper. An email read in part, "The California State Department of Fish and Wildlife at UC Davis will do a necropsy on

'One Eye' as a marker to the disease."

But the gray foxes had already begun to vanish. Something triggered and sent off a tremble through the press. David Johns, cofounder of The Wildlands Project and Yellowstone to Yukon Conservation Initiatives, said, "As we extend the corridors across this nation, be assured that pathogens follow," which tells me that we need to take a long look at ourselves and our science, and try to see ahead. If we create genetically healthy regions, then the overall health of the gray fox will be improved and its own immune system will be better able to fight off the disease. That is a supposition on my part, but one I believe to be well founded.

We need to see and understand nature's intellect. Why should distemper ravage the bush dog population out there in the Baylands? Where did it come from? Where does it live in the wild when not infecting an animal? How can we end it? Perhaps most importantly, if we *can* end something that exists in nature, *should* we?

SIXTY-FOUR

CANINE DISTEMPER

In brief, it was a dark wind that blew through from as far south as we knew down near Sunnyvale and on up to as far north as the Facebook campus. It was a plague. Canine distemper ravaged the populations of gray foxes out there in the Baylands, leading to the total collapse of the ecosystem. The history carved out by all those generations of gray foxes came to an end as death's reverberations surged down the pipeline of the community, affecting every animal within range of that great swath. All changed as nature took another step.

Such a massive death toll battered those who read the article in San Jose's *The Mercury News*, the one that cried an end to an era, to a lineage of foxes that had ruled over a broad expanse of land. I came to know the foxes because they trusted me, because my curiosity drove me, because my need to know stirred my heart even in my sleep.

It's a virus. There are many varieties. This is canine distemper.

Canine distemper (Proceedings)[3]
October 31, 2009
Clinical signs
Distemper virus can invade the respiratory, gastrointestinal, skin, immune and nervous systems. Consequently, signs are highly variable and disease course depends on immune response and dose. Most commonly, early signs of clear to green nasal and ocular discharge, fever, loss of appetite, and depression are seen 1-2 weeks after infection, possibly followed by lower respiratory and gastrointestinal involvement. The "classical" neurological signs usually appear 1-3 weeks after recovery from GI and respiratory disease, but may develop at the same time or months later, even without a prior history of systemic signs.

Clinical signs more suggestive of distemper but seen with less frequency include neurological signs, ocular signs, and dermatological signs. The ocular signs are often a valuable hint that distemper may be the underlying cause of dogs' symptoms. These include: anterior uveitis (inflammation of the front chamber of the eye; may cause the cornea to appear cloudy and/or cause changes in the appearance of the virus); keratoconjunctivitis sicca (dry eye); and optic neuritis (inflammation of the optic nerve-may cause sudden blindness).

Canine distemper virus infects dogs and other mammals, including ferrets and raccoons. Dogs of all ages are susceptible if not previously immunized, although infection is most common in puppies less than 16 weeks of age. Domestic cats are not at risk of distemper, although some large felids such as lions appear to be. (Feline panleukopenia, which sometimes is referred to as feline distemper, is not related to canine distemper).

The article reads, "The ocular signs are often a valuable hint that distemper may be the underlying cause of dogs' symptoms." This is what I have had considerable experience with since 2012, all the way back to when I first noticed Gray's ocular discharge but had no idea what lay behind his eye that could cause such drainage. I guessed that it was a foxtail stuck in the corner of his eye, until I saw both of his eyes become so badly infected that he was blind.

I knew that an infection in both eyes meant that the infectious agent was contagious.

That's when I began checking all of the foxes' eyes on a regular basis. I logged what I saw and reported my findings to anyone who would listen. That was the most obvious recourse, though such knowledge was ultimately useless without the support of those who could have made a difference.

SIXTY-FIVE

VANISHING

I didn't grasp the dynamics of everything staring me in the face. Canine distemper slid into consciousness; I was driven to know more, though I wish I'd never had to learn the things I know now. The distemper destroys the gastrointestinal system through vomiting and diarrhea, invades the neurological system, and, once fully active, courses here and there throughout the body, wrestling for dominance over the animal's ability to survive. The victim becomes lethargic, moving slower and slower until they can no longer summon the strength, develops a wracking cough as their respiratory system gives way to pneumonia, then twitches and convulses in waves of seizures before eventually succumbing to what must be a welcome death.

Gray's immune system waged a heroic war with his canine distemper, but during early November 2016, as already told, the distemper overwhelmed him and he died unable to see. What a magnificent, legendary gray fox he was. The world is lesser without him.

One by one, they vanished over there along Madera Creek: Tippy and Tense disappeared, Blue vanished, and Dark Eyes failed to come at my call. She was my Buddha dog, and I miss her even now.

Dark acted oddly, grew reclusive, then left the area. Some males do that shortly after the pups are raised. I thought that he'd just taken a "vacation," as I called it; Dark seemed to take a "vacation" at every turn. I had no idea that he would slink off into the brush, become too weak to stand, and feel that sinking sensation that would usher him toward unconsciousness forever.

This was Dark, the alpha male of the region, the male that cleared out all of the single males each January. The memory takes me away each time. He stalked Blue, terrorized him away so that they might have their litter in peace, and with no fear of incest. But when I first saw Dark act toward Blue so aggressively, I had no idea why.

One afternoon, Blue sat over on the edge near Marker #17 along the overflow channel. I had passed Dark several minutes before, over on the other side of the ditch. He lay there half awake. I continued on down the channel and welcomed Blue.

"Hey, Blue, how are you?" I called.

Blue sat upright, alert, intense. His demeanor had a twinge of challenge to it. It was as if Blue already knew that Dark was nearby, and the confrontation loomed as Dark came from beneath the willow, stood there at the edge, and stared at Blue, who locked eyes with him.

I watched from the concrete of the overflow channel as it unfolded in what felt like slow motion. An instant before Dark charged, Blue shot through the brush. Dark heard him not far ahead. There was no room for trails, for any rules. Dark and all other alpha male gray foxes had to rid the land of all males, for it seems they had some understanding of incest, even rape, as had happened when Cute, Big Guy's mother, had tried to rise with a growl, and he'd clasped her with his teeth, his body.

Dark had failed to drive his son Big Guy from their home range in time. After several tries, he gave up.

They followed. Just behind me, I heard a fox… Well, you know the story, don't you?

All of them vanished, one by one, and we had no idea why.

Mama Bold vanished. Only in hindsight do I see that she began that river of dying.

In hindsight, I see where I was blind.

When the necropsies were final, I received the following email:

-----Original Message-----

8:05, Tuesday, January 3, 2017

Subject: RE: Gray foxes in Los Palos - follow up, FINAL - Confirmed CDV infection both foxes

Hi all - I have received the final report from the CAHFS Lab at UC Davis.

CDV infection is confirmed as cause of death for both foxes with extensive positive specific staining by immunohistochemistry performed on brain of fox Z16-1315 and lungs of fox Z16-1320.

Additionally, the protozoal cysts are positive for

toxoplasmosis in fox Z16-1315, likely as a consequence to the CDV immunosuppression. (Infection with CDV depresses the immune system, which then can result in latent infections becoming clinically significant or other opportunistic infections taking hold).

All testing is completed at this time.

Given the confirmation, all of my information contained in the email yesterday applies.

Once we have confirmed a CDV outbreak in a localized area in foxes, skunks, or raccoons, we don't have the funds to continue testing every carcass. Instead, I prioritize testing that gives us the most strategic information on trucking the disease.

I plan to test the stored fox that died on November 1 (Peninsula Humane ID 16-3349; WIL ID Z16-1295) - to get a better idea of the time frame (i.e. determine if this older death was CDV also).

After that I can only fund testing of additional carcasses if suspect cases appear in a different area (like the Pacifica animal mentioned), the signs change (a different problem suspected), or mortalities occur in a different species.

We have assisted County/City Animal Services with info dissemination efforts before in other areas, by co-drafting an email or bulletin for local veterinary clinics that alert vets and staff to the issue, provide CDV disease ecology information, and reinforce the prevention and vaccination guidance for pets that I wrote below in yesterday's email.

If that is something that you and Peninsula want to do, I am happy to assist.

Sincerely,

Senior Wildlife Veterinarian (Specialist)

Wildlife Investigations Lab CA Dept. of Fish and Wildlife

This was the first link to the death between us all. Canine distemper was confirmed in both foxes. At long last, there were actionable findings to support that this was indeed a growing problem in the region, one that needed help to solve. We promoted the gray fox, and soon there were a good number of people and organizations backing us every step of the way.

SIXTY-SIX

PUPS IN THE GRASS

I didn't monitor any den as much as I did the one out along Madera Creek. The supervising ranger and I had negotiated my permit, but it was still more restrictive than I would have liked. I knew there were families of foxes out there beyond those imaginary boundaries, because I'd seen signs along the way. It wasn't until 2013 that I finally succeeded in extending my range, and it's a good thing that I did.

I drifted back into Fox Hollow to call for the foxes, and Bold and Gray came from the natal den to where I stood on the road. They remembered my call, the tone, the timbre of my voice. Bold gave me the gray fox happiness dance, pleased to have me there once again.

It had been a long time. As she danced, she slunk her belly low and cocked her ears back and flat along her head, and as she swarmed around me, she moved in a writhing, snakelike fashion. When she reached me, she looked up like she had when she was but a pup on that very same road, and I knew that she remembered and was as happy as I was to see her. I lacked the telling ritual of a human dance that would let her know how I felt. No one had taught it to me.

After a week or so, I wondered why I hadn't seen any pups, or even any signs of them. I mean, after all, the buckthorn berry season had just passed. Elsewhere, the gray fox pups had already devoured the deep purple berries and seeds, resulting in patch upon patch of smelly scat fields that would grow into large purple latrines on the road and everywhere else.

In Gray and Bold's territory around Fox Hollow, scat was nowhere to be found, and the buckthorn berries were everywhere. There were no signs of any pups, no scat, so I wondered, *Is she barren?* I doubted whether Gray would hang with her if she were barren, but maybe that didn't matter to him. There had to be something else.

It was obvious that the two pups back on the channel were sick. They were infected with pus. One Eye was blind in his right eye, and half-blind in the other. Then he vanished.

We were still suffering under the shrouds of ignorance. I told myself lies about why Mama Bold had vanished, why two of her pups were gone, until I ran out of tall tales. Slowly, I accepted that Mama Bold had died, and then I began to see that everything in Fox Hollow was unstable. I never knew how many pups I might see on a given day, but the numbers were dwindling. Even along Madera Creek, Dark had been gone for more than a month. Life among the adults back there worked as it should, except for the two pups lying there out in the grass, totally without cover in the hot sun. They hardly twitched when I walked up beside them, and I knew that something was terribly wrong.

Death of Silicon Valley Gray Foxes Point to Urgent Need for Wildlife Corridors

BY TOM MOLANPHY – JANUARY 13, 2017

DNA bottlenecks, inbreeding make animals susceptible to disease, says "Fox Guy"

Bill "Fox Guy" Leikam hopes the most recent chapter in the story of the Silicon Valley urban fox is not the beginning of a tragedy for connecting healthy ecosystems. As reported by the *San Jose Mercury News* last week, up to 18 urban gray foxes belonging to four different "skulks," or groups, that Leikam has studied and researched over the last seven years in Los Palos, California, died last month of canine distemper – a contagious viral infection with no known cure.

This is the first time in recent memory that local wildlife observers have seen such a big wildlife die-off. "We have 12 fox carcasses and six more that are missing and presumed dead," Leikam, founder of the Urban Wildlife Research Project, told *Earth Island Journal*. "[December] was like a dark wind carrying the virus as it swept through, taking all of the foxes throughout the region and possibly even up into East Los Palos."

The gray fox, a small, tree-climbing member of the canid family, is one of the few wild carnivore species that seems to have successfully adapted to living in and around big cities, though it still faces many threats. A small population of these urban-dwelling canids, comprising several skulks, have captured the hearts of residents and researchers in California's Silicon Valley, including Leikam who has been researching their role in the local ecosystem as well as the challenges urban habitats present to grey foxes.

It's hard to pin down the exact number of foxes living in the South Bay Area. Leikam says they seem to be living in "pockets" and regions from south Redwood City, south through Alviso and up the eastern side of the Bay to at least the southern edge of the Oakland International Airport. They have also been spotted in the foothills of Los Altos, Saratoga, and on south.

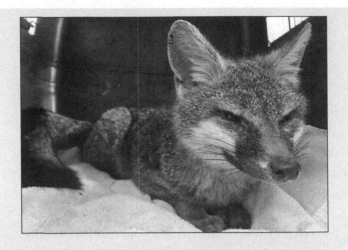

Watching animals die from distemper – especially animals you have studied and protected and know by name – is not something "I ever want to live through again," said the retired schoolteacher whose dogged monitoring of the foxes earned him the "Fox Guy" moniker.

Unlike animals with rabies who turn aggressive, animals with canine distemper turn incredibly lethargic. "They won't even move out of way," Leikam said. "The distemper attacks their respiratory, nervous, immune and gastrointestinal systems and their brains swell. During the last stage before their systems collapse, which has been documented, they walk around as if they're drunk... and within half an hour they're dead."

All roads from this local catastrophe lead to the urgent need for nature corridors as well as the overall understanding of ecosystem connectivity, said Leikam, who hopes to eventually help create a comprehensive "San Francisco Bay Area Wildlife Corridor" that would protect this region's rich natural heritage. Leikam hopes the decimation of this urban fox population can vault the subject of nature corridors into more mainstream circles. "It's all happening in our backyards. We just need to be a little more attentive about what is happening with the wildlife around us."

The basic idea for wildlife corridors has been around for some time, most noticeably – and dramatically – promoted by renowned scientist E.O. Wilson and his notion of a "Half-Earth," i.e., the need to dedicate a full half of Earth to natural systems in order to help species survive and thrive. One of the most prominent organization promoting ecosystem connectivity is the Yellowstone to Yukon Conservation Initiative, the nonprofit that has been working to connect Yellowstone to Yukon since 1997.

"It makes sense that when animals are in an 'island' and inbreeding occurs, that the genetic pool becomes degraded and the immune system grows weak," Leikam said. "It's easier for distemper to attack. But once you open up the corridor, that increases genetic diversity, which in turn increases the body's ability to fight disease."

Although the range of the Los Palos urban foxes is much smaller than the area encompassed by the Yellowstone to Yukon proposal, the principles remain the same – wild animals need to able to travel between habitats. One big difference, as any living creature in the Bay Area knows, is that this space is coveted by many different interests. Put the needs of a small canine up against the financial interests of a billion-dollar developer, and you can start to understand the battles Leikam has faced over the years.

David Johns of the Wildlands Network, a conservation group that's pushing for the creation of a network of wildlife corridors or "wildways" across North America, has been an avid supporter of nature corridors and overall connectivity for some time. "The important larger message here is the loss of habitat," Johns said. "These foxes are on the edge and it doesn't take much to push them over that edge."

Add into that the real possibility of a federal government that will be, at best, totally disinterested in urban wildlife and, at worst, hell-bent on defunding every policy that has the word "environmental" anywhere near it, and it's not hard to understand the depths of Leikam's concern. "It's all speculation at this point... but some of the [federal] cabinet appointees to some of the conservation areas seem to back oil companies more than conservation," he said.

Both Leikam and Johns are part of a long environmental movement that have asked simple but essential questions, such as: What if the policies, laws and guidelines that regulate human behavior were based on long-term, environmentally sustainable stewardship instead of short-term, profit-driven projects?

For now, Leikam plans to press on. He did spot a few "skittish" foxes last week – young, first year, gray foxes dispersing from their birth areas to establish their home range and find a mate. Leikam even checked in with the Facebook campus in Menlo Park. A family of foxes that has adopted the campus as home and has taken on rock-star status, and was happy to find out that "the foxes there look as normal as can be."

But the long-term sustainability of the urban fox will involve a coordinated effort to promote – and a broad understanding to accept – the notion that such wildlife not only poses no threat to humans, but that they can provide benefits such as rodent control, as well.

"What we need most is funding for our DNA project," Leikam said. "That project will help us identify relationships between the urban foxes and find out exactly where the inbreeding has happened. Since this population has been wiped out, we'll have to move into new areas to do that DNA work, as well as continue our study at the Los Palos Baylands Preserve," he said. "We're moving forward."

Earth   Island   Journal: (https://www.
earthisland.org/journal/index.php/articles/
entry/death_of_silicon_valley_grey_foxes_
point_to_urgent_need_for_wildlife_corrid/)

Photo 1: A dead gray fox pup. In December,
18 gray foxes belonging to four different
skulks in and around Los Palos died of canine
distemper.

Photo 2: The last living image of a pup
called One Eye, who too, succumbed to the
disease. Animals dying from distemper is
not something Leikam ever wants to witness
again.

SIXTY-SEVEN

WILDLIFE DETECTIVE

After the die-out of the foxes at the Los Palos Nature Refuge, people often asked whether I knew how the canine distemper arose and infected so many foxes. I had no idea until recently; I may have hit upon at least one possible answer, yet this is still conjecture. People asked, "How long has distemper been around?" I emailed the lead veterinarian and asked. Her assistant replied that it couldn't have been more than a month.

I didn't think that could be true, so I went to work on trying to find the answer and made a discovery on Friday, March 24, 2017. It was one of those "aha" moments when the right piece of information falls exactly into place. When I found where it fit, the rest of the puzzle was illuminated so that I could look deeper and understand something I had never before known. Along with that discovery rode the excitement of uncovering and deciphering the unknown, albeit fettered by the sorrow at hand.

I happened upon this new data because I needed to refine my investigation of the gray fox's ocular drainage, one of the overt symptoms that points toward infection from viral canine distemper. I needed to know when and where in my logs I had entered the first account of seeing Gray with his eyes draining an ugly fluid that I'd taken to be just a passing infection. I guessed that it was sometime in 2014. I needed to be sure. I went into my gray fox log and punched in a series of keywords.

Gray didn't show eye drainage until June 22, 2015. I thought, *That can't be right. It has to be much earlier than that.* I ran several searches using several keywords, but always had the same result. I kept looking for something in 2014, but nothing emerged to change the results of my findings. I was baffled.

But then I remembered that part one of the log ended on April 2, 2013, the day my hard drive had crashed.

I ran a search through the first log and bingo: Mama Bold had had an eye infection as far back as July 30, 2013. In those early years, ocular discharge had been a mystery. I copied the date and entered it into a search, and a picture of Mama Bold surfaced. It was exactly the kind of eye drainage seen in almost every fox with canine distemper, but back then, we had had no idea what canine distemper looked like. Gray's infection didn't show until two years later.

That finding changed everything, at least regarding my thoughts on the distemper. It showed that the lead vet assistant had been wrong. People had been asking: "Where did the distemper come from?" Often, I'd reply that no one knew, but it could have been due to the high population numbers. Elsewhere, we'd heard that when a gray fox population saturates a region, a die-off sometimes occurs, but no one knew why.

The state's lead vet said that the canine distemper that decimated a total of twenty-five foxes was endemic to California. The foxes that grew up there were, in part, a product of inbreeding, and as such their immune systems controlled the virus until they were so deeply compromised that they could not survive. This condition spread over a broad territory, from the banks of Arroyo Creek in the north to Hard Rock in the south.

This was an act of nature, but one so heavily influenced by unseen factors that it was simply lost in a sea of catalysts and causes. There was no single root cause, and there is no method of correcting what was done in the past. We can only look forward.

As for me, I'll never stop calling them.

SIXTY-EIGHT

AN ENDING

Over the years, Mama Bold and Gray had only three litters that made it to maturity. Although the pups of 2016 experienced marginal survival for a while, all five eventually died. No predators were involved there, as I sometimes suspected. They all died due to canine distemper passed on to them by the prior generation: Mama Bold and Papa Gray, Dark and Cute, the enclave of foxes along Madera Creek, and the nameless foxes at the golf course.

The implications of this situation are ominous, and only clarify that we must get this study underway now. We need to know what viruses are in the area, how widespread they are, and eventually how to eradicate them so that the gray foxes, raccoons, and other wildlife do not succumb to canine distemper, or any other disease, for that matter. That is our directive as humans, and I believe it is well within our capacity to accomplish all this and more.

Though they vanished, I continued to call. I hoped that maybe, just *maybe*, they had gone off somewhere and would come back. I'd seen that happen before, but one by one, and sometimes more, they quit coming from the brush, quit lazing in their regular places near Marker #17, or up on the hill next to Sambeek's Shed.

All of the foxes succumbed to a disease beyond their control or ken, but I was in the dark. I didn't know better, and so intense hope drove me onward. In some ways, it still does. I called them and—

I called for the foxes by name.
"Mama Bold. Hey, Mama, you here?"
But she didn't come from the brush.
I called, "Hey, Dark, hey there.
Whatcha doin'? Hey, little fox,"
And I rolled the word "little"
just for him.
They knew me before I called;
They smelled me

And heard the gait of my boots
In the deep bone grass.
I called,
"Hey, little fox. Hey, where ya going?
Hey, Daring.
Hey, Pale and Big Guy.
Hey, Blackie.
Hey, little fox." I called
And called, "Hey, little fox,"
But none came
From under the old coyote bush,
Nor did the pups
Run-tumble toward me
In the early morning
From on up the road
As I came along.
Brownie didn't show
From across the marsh
But I kept calling
And calling again
And again,
"Hey, Brownie, you over there?"
Please come
And surprise me.
"Hey, Helper, come on over."
Just let one dash

From under the trees' canopy.

Just one.

"Hey, little fox," and I rolled

The word "little"

Just for me.

"Hey, Cute.

Hey, sisters Tippy and Tense.

Hey, Dark Eyes, my Buddha dog."

And I called for endless days.

Trying to resurrect

My precious gray fox,

I called.

ENDNOTES

1. Mitochondrial Analysis of the Most Basal Canid Reveals Deep Divergence Between Eastern and Western North American Gray Foxes (Urocyon spp.) and Ancient Roots in Pleistocene California, Natalie S. Goddard, Mark J. Statham, Benjamin N. Sacks - PLOS | One, August 19, 2015

2. http://www.native-languages.org/gitchi-manitou. htm

3. From DVM360: https://www.dvm360.com/view/canine-distemper-proceedings-0

ABOUT W.C. LEIKAM

From October 2009 to the present, Bill Leikam, President & CEO has conducted unprecedented, groundbreaking field research on the behavior of the gray fox (Urocyon cinereoargenteus). People know Bill as the Fox Guy. He is on the Board of Directors for Guadalupe-Coyote Resource Conservation District. Bill has many accomplishments to his name including being a published live jazz reviewer for All About Jazz, contributed to the field guide *Canids of the World* by Dr. Jose Castello, published by the Princeton University Press, been written about in Beth Pratt's book *When Mountain Lions are Neighbors*, and has been written about in varied magazines and news articles. In 1981 he was a Delegate to the People's Republic of China based on his research into the nature of consciousness.